THE ILLUSTRATED FLORA OF ILLINOIS

The Illustrated Flora of Illinois

ROBERT H. MOHLENBROCK, General Editor

ADVISORY BOARD:

Constantine J. Alexopoulos, *University of Texas*
Gerald W. Prescott, *University of Montana*
Aaron J. Sharp, *University of Tennessee*
Floyd Swink, *The Morton Arboretum*
Robert F. Thorne, *Rancho Santa Ana Botanical Garden*
Rolla M. Tryon, Jr., *The Gray Herbarium*

THE ILLUSTRATED FLORA OF ILLINOIS

GRASSES
panicum to danthonia

Robert H. Mohlenbrock

SOUTHERN ILLINOIS UNIVERSITY PRESS
Carbondale and Edwardsville

FEFFER & SIMONS, INC.
London and Amsterdam

QK
495
.674
M563

Library of Congress Cataloging in Publication Data

Mohlenbrock, Robert H
  Grasses: Panicum to Danthonia.

  (The Illustrated flora of Illinois)
  Bibliography: p.
  1. Grasses—Illinois—Identification. I. Title.
II. Series.
QK495.G74M563   584'.9'09773   73–6807
  ISBN 0–8093–0521–6

Copyright © 1973 by Southern Illinois University Press
All rights reserved
Printed in the United States of America
Designed by Andor Braun
Illustrations by Miriam W. Meyer

This book is dedicated to
Dr. George B. van Schaack,
devoted bibliophile and student of grasses

CONTENTS

ILLUSTRATIONS

FOREWORD

The grasses of Illinois will appear in two volumes of The Illustrated Flora of Illinois series. It follows publication of the ferns of Illinois and two volumes on monocotyledonous plants of Illinois. The series grew out of an idea to present all information known about every kind of plant which occurs in Illinois. The Illustrated Flora of Illinois is a multivolumed flora of the state of Illinois, to cover every group of plants, from algae and fungi through flowering plants. In addition to keys and descriptions of every plant known to occur in Illinois, there would be provided illustrations showing the diagnostic characters of each species.

An advisory board was set up in 1964 to screen, criticize, and make suggestions for each volume of The Illustrated Flora of Illinois during its preparation. The board is composed of taxonomists eminent in their area of specialty—Dr. Gerald W. Prescott, University of Montana (algae); Dr. Constantine J. Alexopoulos, University of Texas (fungi); Dr. Aaron J. Sharp, University of Tennessee (bryophytes); Dr. Rolla M. Tryon, Jr., The Gray Herbarium (ferns); Dr. Robert F. Thorne, Rancho Santa Ana Botanical Gardens and Mr. Floyd Swink, the Morton Arboretum (flowering plants).

This author is editor of the series and will prepare many of the volumes. Specialists in various groups are preparing the sections of their special interest.

There is no definite sequence for publication of The Illustrated Flora of Illinois. Rather, volumes will appear as they are completed.

Robert H. Mohlenbrock

Southern Illinois University
December 1971

The Illustrated Flora of Illinois

GRASSES

panicum to danthonia

County Map of Illinois

Introduction

The nomenclature for species followed in this volume is based largely on that of Hitchcock (1950) in the *Manual of the Grasses of the United States,* except where recent monographs and revisions are available. The division of the grass family into subfamilies and tribes essentially follows Gould (1968) and is a major departure from the sequence usually found in most floristic works in North America.

Synonyms, with complete author citation, which have applied to species in the northeastern United States, are given under each species. A description, based primarily on Illinois material, is provided for each species. The description, while not necessarily intended to be complete, covers the more important features of the species.

The common name (or names) is the one used locally in Illinois. The habitat designation is not always the habitat throughout the range of the species, but only for it in Illinois. The overall range for each species is given from the northeastern to the northwestern extremities, south to the southwestern limit, then eastward to the southeastern limit. The range has been compiled from various sources, including examination of herbarium material. A general statement is given concerning the range of each species in Illinois. Dot maps showing county distribution of each grass in Illinois are provided. Each dot represents a voucher specimen deposited in some herbarium. There has been no attempt to locate each dot with reference to the actual locality within each county.

The distribution has been compiled from field study as well as herbarium study. Herbaria from which specimens have been studied are the Field Museum of Natural History, Eastern Illinois University, the Gray Herbarium of Harvard University, Illinois Natural History Survey, Illinois State Museum, Missouri Botanical Garden, New York Botanical Garden, Southern Illinois University, the United States National Herbarium, the University of Illinois, and Western Illinois University. In addition, a few private collections have been examined.

Each species is illustrated, showing the habit as well as some

1

of the distinguishing features in detail. Most of the illustrations have been prepared by Mirian Wysong Meyer. Dr. Kenneth Lewis Weik illustrated 208, 211, 212, 213, 214, 217, 219, 220, 221, 222, 223, 224, 278, 284, and 285. Fredda Burton prepared figures 4, 9, 10, 21, 22, 26, 79, 80, 91, and 102.

Several persons have given invaluable assistance in this study. Mr. Floyd Swink of the Morton Arboretum has read and commented on the entire manuscript. For courtesies extended in their respective herbaria, the author is indebted to Dr. Robert A. Evers, Illinois Natural History Survey; the late Dr. G. Neville Jones, University of Illinois; Dr. Glen S. Winterringer, Illinois State Museum; Dr. Arthur Cronquist, New York Botanical Garden; Dr. Jason Swallen, the United States National Herbarium; Dr. Loren I. Nevling, the Gray Herbarium; Dr. George B. van Schaack, formerly of the Missouri Botanical Garden and the Morton Arboretum; and Dr. Walter Lewis, the Missouri Botanical Garden.

Southern Illinois University provided time and space for the preparation of this work. The Graduate School and the Mississippi Valley Investigations and its director, the late Dr. Charles Colby, all of the Southern Illinois University, furnished funds for the field work and the salaries for the illustrators.

HISTORY OF GRASS COLLECTING IN ILLINOIS

There always have been many kinds of grasses known from Illinois. Mead, who published the first extensive list of Illinois plants in 1846, recorded 83 species of grasses. Lapham, who wrote specifically about Illinois grasses in 1857, listed some 135 species. Patterson, in his catalog of Illinois plants nineteen years later (1876), reduced this number to 123 species.

The most important work on Illinois grasses has been by Mosher, in 1918, when she prepared the *Grasses of Illinois*, a treatise providing descriptions and cited specimens of all grasses known to occur in the state. Two hundred and four species are recorded in her work.

Jones reported 212 species of grasses in 1945 and 220 in 1950. Jones, Fuller, Winterringer, Ahles, and Flynn added 26 species in 1955, bringing the total to 246. In 1960, Winterringer and Evers included 8 additional species of grasses from Illinois. Glassman has studied the grasses of the Chicago region thoroughly for the past several years, and his treatment of these (1964) is excellent. During the research for this book, several species of grasses previously unreported from Illinois were found in various herbaria, for the most part bearing misidentifications. A number of additional species was found during intensive field study, particularly in the southern one-third of the state. Differences in the taxonomic treatment have accounted for the addition or subtraction of some species within the state. In these volumes on grasses 286 species are recognized from Illinois, along with 49 lesser taxa.

MORPHOLOGY OF GRASSES

Grasses belong to the family Poaceae (also called Gramineae). Until recently, most botanists grouped grasses and sedges (Cyperaceae) in the order Graminales (or Poales). Anatomical, morphological, and other more recent evidence show that, in addition to grasses and sedges, other families such as the Xyridaceae, Commelinaceae, Pontederiaceae, and Juncaceae share some of the same characters and should be grouped together. This view is followed here so that these six families are considered to comprise the Commelinales. The Xyridaceae, Commelinaceae, Pontederiaceae, and Juncaceae are treated in *Flowering Plants: Flowering Rush to Rushes* (1970); the Cyperaceae will be forthcoming in two subsequent volumes.

The nature of grass structures generally is so different from that of other flowering plants that a special terminology is applied to grasses. A thorough understanding of these terms will enable one to identify more readily an unknown specimen.

Grasses are annuals biennials, or perennials. Annuals have tufts of fibrous roots and live for a single growing season. Perennials may be tufted (*Fig. 1*), or they may have rhizomes (horizontal, root-producing stems below ground [*Fig. 2*]), or they may have stolons (horizontal, root-producing stems above ground [*Fig. 3*]), or a short, thick, subterranean crown (*Fig. 4*).

The stem which bears the leaves and inflorescence is called the culm. While the culm may be hollow or solid, the nodes (where the leaves arise) are nearly always solid. The culms may be simple or branched. Often they are jointed (geniculate) near the base (*Fig. 5*). Culms may be erect, divergent (spreading), or prostrate and matted.

Grass leaves are borne at the nodes in two planes along the culm. This condition is referred to as 2-ranked (*Fig. 6*). Some-

times, because of a twisting of the culm, the 2-ranked condition
is not apparent. The leaf is composed of a blade and a sheath.
The sheath wraps around and encloses a portion of the culm. If
the margins of the sheath are united, forming a cylinder, the
sheath is closed (*Fig. 7*); if the margins are not united, the
sheath is open (*Fig. 8*). The blade is the free portion of the leaf.
It is parallel-veined and generally elongated, although some
grasses with rather short, broad blades occur. The blades normally are flat, but they may be folded (plicate [*Fig. 9*]) or inrolled into a slender tube (involute [*Fig. 10*]). Along the inner
face of the leaf, where the blade adjoins the sheath, there is often
a ciliate, membranous, or cartilaginous structure of varying size
and shape known as a ligule (*Fig. 11*). In some grasses, some of
the leaves are not blade-bearing, therefore consisting merely of
sheaths.

The inflorescence is the aggregation of a group of spikelets
(the basic unit of the grass inflorescence). An elongated, simple
axis with pedicellate spikelets borne along it is called a raceme
(*Fig. 12*); if the spikelets are sessile along the simple axis, the

1. Tufted perennial.   2. Rhizome.

3. Stolon.   4. Short, thick,   5. Geniculate
             subterranean crown.   base of stem.

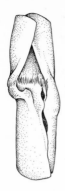

6. Two-ranked leaves.

7. Closed sheath.

8. Open sheath.

9. Plicate leaf.

10. Involute leaf.

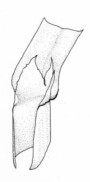

11. Ligule.

12. Raceme.

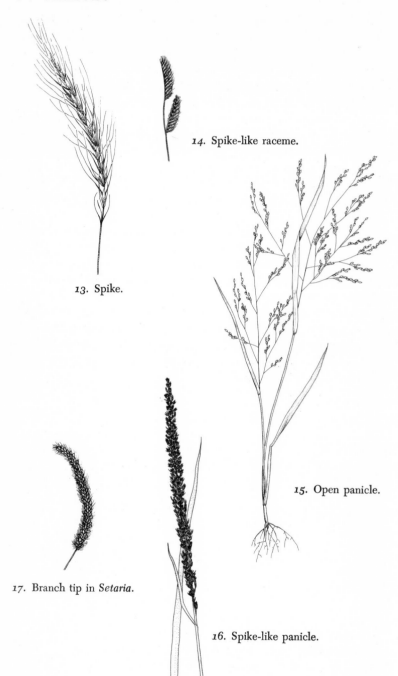

14. Spike-like raceme.

13. Spike.

15. Open panicle.

17. Branch tip in *Setaria*.

16. Spike-like panicle.

inflorescence is a spike (*Fig. 13*). Short-pedicellate spikelets crowded on an elongated, simple axis make up the spike-like raceme (*Fig. 14*). If the inflorescence is branched, and the spikelets are pedicellate, the term used is panicle (*Fig. 15*). The panicle may be very wide-spreading and open (diffuse [*Fig. 15*]), or it may be contracted so much as to resemble a spike (*Fig. 16*). This latter situation gives rise to the term spike-like panicle.

The tip of each branch of the panicle normally bears a spikelet, although in *Setaria* (*Fig. 17*) and *Cenchrus* (*Fig. 18*), some of the branch tips are sterile and modified into bristles.

The spikelet is composed of an axis, called the rachilla, along which are borne bracts in two ranks (*Fig. 19*). The lowest two bracts bear no flowers in their axils. These "empty" bracts are the glumes. They frequently are unequal in size although rarely unlike in texture. Both glumes are essentially lacking in *Leersia* and *Zizania*, while the first (lower) glume is usually absent in *Paspalum, Digitaria, Eriochloa,* and *Lolium. Elymus hystrix* usually has its glumes reduced to awns. A sharp ridge down the back of a compressed glume is called the keel (*Fig. 20*). Sometimes the entire spikelet falls at maturity, while in other species the glumes remain behind. In the first case, the spikelet is said to disarticulate below the glumes, while in the latter case, it is said to disarticulate above the glumes.

18. Branch tip in *Cenchrus.*

19. Spikelet with bracts   20. Keel on glume.
in two ranks.

Above the glumes are one or more bracts which usually bear a flower within. These fertile bracts are the lemmas. Facing each lemma is a usually somewhat smaller palea (*Fig. 21*). Between the lemma and the palea is the flower (*Fig. 21*). In *Chasman-*

thium and *Panicum,* the lowest lemma does not produce a flower, while in *Melica* and *Chloris,* the uppermost lemma is sterile. In *Phalaris,* the two lowest lemmas are reduced to scales. Lemmas generally are of the same texture as the glumes, although the fertile lemma in *Panicum* is indurated. The callus of a lemma may refer to a swollen, hardened area at its base (as in *Stipa* and *Aristida* [*Fig. 22*]) or a tuft of hairs (as in *Calamagrostis* [*Fig. 23*]). Lemmas also may be keeled. Spikelets with a single fertile lemma are said to be 1-flowered (*Fig. 24*), while those with two or more fertile lemmas are several-flowered (*Fig. 25*).

The palea is smaller than the lemma and usually of more

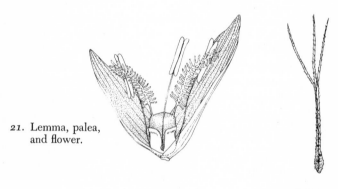

21. Lemma, palea, and flower.

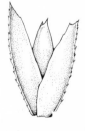

22. Callus at base of lemma.

23. Tuft of hairs at base of lemma.

24. One-flowered spikelet.

delicate texture. In *Panicum hians,* the palea becomes indurated at maturity. The palea is often absent in *Agrostis.* Most paleas have two keels down the back.

The grass flower is much reduced from the flower of Liliaceae and other more showy flowering plants. It consists of three stamens (occasionally 1–6) and one pistil. Each stamen bears a 2-celled anther. Each pistil is 1-celled, with but one ovule, but there usually are 2–3 styles. At the base of the flower usually are found 2–3 small scales thought to represent the perianth. These scales are the lodicules (*Fig. 26*).

26. Lodicules.

25. Several-flowered spikelet.

Most grasses have a fruit known as a caryopsis, or grain. The seed-coat of the single seed is united directly to the matured ovary wall (pericarp). (The pericarp is free from the seed in *Eleusine, Crypsis,* and *Sporobolus*). At maturity, the grain drops free from the lemma and palea, or it may fall while enclosed by the lemma and palea.

The lemma, palea, and enclosed flower comprise the floret.

DISTRIBUTION OF ILLINOIS GRASSES

Grasses occur in every possible habitat in Illinois—from standing water to the driest bluff-tops, from prairies to woodlands, from waste places and fields to the deepest canyons. Most species of grasses are restricted to a basic type of habitat. A few, such as *Poa pratensis, Agrostis alba,* and *Setaria lutescens,* may be found virtually in every habitat. The following discussion of habitats for Illinois grasses is divided into three major sections: moist, dry, and waste areas.

Moist Habitats

STANDING WATER   There are few grasses, indeed, which can tolerate partial submergence in water. Those which do occupy this kind of habitat are not common and are very locally distributed. Probably the most widespread aquatic grasses in Illinois are *Glyceria septentrionalis* and *Alopecurus aequalis*. *Zizania aquatica* is found in the northern two-thirds of Illinois, while *Deschampsia flexuosa* is known only from extreme northeastern counties. Predominantly southern species of aquatic grasses are *Paspalum fluitans,* restricted to the southern two-thirds of the state, and *Glyceria arkansana* and *Puccinellia pallida*, which are confined to a single station in the southern tip of Illinois.

MOIST SOIL   In this paragraph will be considered grasses which grow in moist soil, but not generally in woodlands, prairies, or on sandy shores. These are the grasses of low meadows and thickets. The most widespread of these species is *Glyceria striata*, although several others are found locally throughout Illinois. Numbered among these are *Eragrostis hypnoides, Alopecurus carolinianus, Muhlenbergia racemosa, Leersia lenticularis, L. oryzoides, Echinochloa walteri, E. pungens,* and *Panicum lanuginosum* and *P. clandestinum.* These species usually occur in considerable abundance where they are found. A few of the moist-meadow species are more common in northern Illinois. These are *Phragmites australis, Poa palustris, Calamagrostis canadensis,* and *Agrostis alba* var. *palustris.* Other species, such as *Chasmanthium latifolium, Paspalum pubiflorum, P. laeve, Panicum rigidulum,* and *P. anceps* are principally southern. *Arundinaria gigantea* often forms dense thickets (canebrakes) in lowlands in the southern one-fourth of the state.

MOIST SAND   On the sandy shores adjacent to the major waterways of Illinois are a few characteristic grasses. Although several species occasionally occupy this habitat, the most typical are *Eragrostis frankii, Paspalum ciliatifolium, Cenchrus longispinus,* and *Leptochloa filiformis.* These are grasses which, for the most part, can tolerate the wave action of the larger rivers.

WET PRAIRIES   Wet prairies may be regarded as low, moist,

treeless areas with predominantly prairie vegetation. *Spartina pectinata* indicates this type of habitat, although *Sphenopholis obtusata* var. *major* is usually present as well. *Panicum lanuginosum* var. *implicatum* may be found here with some regularity, particularly in the northern counties.

MOIST WOODLANDS This habitat may occur in a low, rather open terrain, or it may be in the depths of picturesque canyons and ravines. In some cases, scattered boulders may be strewn across the forest floor. A few species seemingly need the protective presence of these boulders for their survival. Species such as *Melica mutica, M. nitens, Muhlenbergia tenuiflora,* and *M. sylvatica* apparently survive better in the moist, rocky woods. Other species are less dependent on the rocks and grow well in essentially rockless woods. Over a dozen species occur rather commonly in moist woodlands throughout the state. These are *Bromus ciliatus, B. pubescens, Festuca obtusa, Poa sylvestris, Elymus hystrix, Sphenopholis obtusata* var. *major, Cinna arundinacea, Muhlenbergia frondosa, M. mexicana, Agrostis hyemalis, Brachyelytrum erectum,* and *Leersia virginica.* These species frequently occur singly and rarely form extensive patches. One species, characteristic of many moist woodlands in northern Illinois, is *Bromus purgans.* Species confined to moist woods in the southern two-thirds of Illinois are *Muhlenbergia bushii, M. glabrifloris, Panicum microcarpon, P. polyanthes,* and *P. boscii.*

Dry Habitats (excluding fields)

ROCK LEDGES Exposed rock ledges, frequently becoming intensively xeric during midsummer, nonetheless may be suitable for the growth of a limited number of species, including some grasses. Characteristic of these xeric ledges are *Vulpia octoflora, Agrostis elliottiana, Danthonia spicata, Panicum gattingeri, Sporobolus vaginiflorus,* and *S. neglectus.* In the southern tip of the state *Andropogon virginicus* occurs along these ledges.

DRY SAND Two distinct areas in which the plants grow in dry sand are found in the northern half of the state. Most extreme is the sand of the dunes along Lake Michigan. The characteristic grasses of this rugged habitat are *Ammophila breviligulata* and *Calamovilfa longifolia.* The other sand habitat is the sandy prairie, such as those studied extensively by Gleason in 1910 in the Hanover, Dixon, and Havana areas. Grasses are a vital component of these sandy areas, serving as sand binders in most instances. Characteristic sand-prairie taxa are *Eragrostis trichodes, Stipa spartea, Leptoloma cognatum, Panicum villosissimum* var. *pseudopubescens, Aristida tuberculosa, Sporobolus cryptandrus, Triplasis purpurea, Koeleria macrantha, Schizachyrium scoparium,* and *Andropogon gerardii.*

DRY PRAIRIES The prairies considered in this paragraph are those found neither in low, moist situations nor in sandy areas. They are of two basic types in Illinois, being situated atop predominantly limestone bluffs or on glacial till (hill prairies), or on generally flat terrain. In either case, the same species usually prevail. Common taxa throughout the state are *Schizachyrium scoparium, Andropogon gerardii, Sorghastrum nutans, Koeleria macrantha, Sporobolus heterolepis, Panicum praecocius,* and *P. oligosanthes* var. *scribnerianum.* Of more limited distribution are *Stipa spartea, Bouteloua curtipendula, Panicum perlongum,* and *Andropogon virginicus.*

DRY WOODLANDS As with species of the moist woodlands, there are some species which seemingly thrive better when boulders are present in the woodlands. Those dry, rocky woodland species most characteristic are *Eragrostis capillaris, Muhlenbergia sobolifera,* and *Panicum latifolium.* In the southern one-third of the state, *Panicum dichotomum* var. *barbulatum* becomes an important species of this habitat. In the dry, essentially rockless woodlands, several taxa regularly may be found throughout the state. Included among these are *Elymus canadensis, E. villosus, E. virginicus, Sphenopholis obtusata, Danthonia spicata, Aristida oligantha, Agrostis perennans, Panicum depauperatum, P. lanuginosum* var. *lindheimeri, P. villosissimum,* and *Sorghastrum nutans.* Confined to northern Illinois, but characteristic, is *Bromus kalmii,* while restricted to southern Illinois are *Panicum laxiflorum, P. dichotomum,* and *Andropogon virginicus.*

Fields and Waste Ground

Grasses, more than any other plants, seem to have the ability to come into and establish themselves in fields, waste ground, and other open areas. Most grasses which occupy this habitat would be considered weedy. A surprising number of these is native. In the following lists, only those grasses which are common throughout most of the state are considered.

Native species

Agrostis alba
Agrostis hiemalis
Aristida oligantha
Aristida ramosissima
Elymus canadensis
Elymus virginicus
Eragrostis pectinacea
Hordeum pusillum
Muhlenbergia schreberi

Panicum capillare
Panicum dichotomiflorum
Panicum virgatum
Poa chapmaniana
Schizachyrium scoparium
Setaria lutescens
Sorghastrum nutans
Tridens flavus

Adventive species

Agropyron repens
Bromus commutatus
Bromus inermis
Bromus racemosus
Bromus secalinus
Bromus tectorum
Dactylis glomerata
Digitaria ischaemum
Digitaria sanguinalis
Echinochloa pungens
Eleusine indica
Eragrostis cilianensis

Eragrostis poaeoides
Festuca pratensis
Hordeum jubatum
Lolium multiflorum
Lolium perenne
Phleum pratense
Poa annua
Poa compressa
Poa pratensis
Setaria faberi
Setaria viridis
Sorghum halepense

USEFULNESS OF GRASSES

Grasses are undoubtedly the most valuable plants to mankind. Grasses used for food by man and his domesticated animals are many. Man utilizes such grasses as barley, corn, millet, oats, rice, rye, sorghum, sugar cane, and wheat in his own diet. Many forage grasses are used for hay, silage, and pasturing. In the open expanses of the western United States, pasture grasses are referred to as range grasses. Many of these are important members of the grass flora of Illinois. Grains of many grasses are important in the diet of wildlife and fowl.

Man employs other grasses for his benefit and enjoyment. Grasses with strong rhizomes are an important tool against soil erosion. In the dunal region of Lake Michigan, certain grasses are valuable sandbinders. Lawn grasses become more and more important to modern living. Several attractive grasses are cultivated for their ornamental value.

RELATIONSHIP OF THE GRASSES

Grasses (Poaceae) and sedges (Cyperaceae) have long been placed near each other in most phylogenetic schemes. Indeed, these two groups share several similar characters – general habit, reduced flowers subtended by an assortment of scales or bracts, similar leaves, 1-seeded fruits, etc. Near to these families, Hutchinson (1959) and others have placed the rushes (Juncaceae) which, although similar in general habit and the presence of inconspicuous flowers, possess an actual perianth.

Recent evidence seems to point to a relationship of grasses, sedges, and rushes to several other families. In particular, grasses appear to be closely related to the Flagellariaceae (of Old World tropics and subtropics), the Restionaceae (of the Southern Hemisphere and Indochina), and the Centrolepidaceae (of the Southern Hemisphere).

HOW TO IDENTIFY A GRASS

Beginning on page 15 is a key for the identification of the genera of the grasses of Illinois. A botanical key is a device which, when properly employed, enables the user to identify correctly the plant in question. It is the intent of this key to use characters which are easy to observe and to avoid the more technical characters which often best show relationships.

Once the genus is ascertained by using the general key, the reader should turn to that genus and use the key provided to the species of that genus if more than one species occurs in Illinois. Of course, if the genus is recognized at sight, then the general keys should be by-passed.

The keys in this work are dichotomous, i.e., with pairs of contrasting statements. Always begin by reading both members of the first pair of statements. By choosing that statement which best fits the specimen to be identified, the reader will be guided to the next proper pair of statements. Eventually, a name will be derived.

Illustrated Key to the GENERA of Grasses in Illinois

* THE ASTERISK FOLLOWING THE NAMES OF CERTAIN GENERA INDI-
CATES THAT THESE GENERA ARE TO BE FOUND IN THE FIRST VOLUME
OF GRASSES IN THIS SERIES.

1. Culms woody_____80. **Arundinaria,** p. 337
1. Culms herbaceous.
    2. Spikelets enclosed by a spiny bur (**Fig. 27**)_____
    _____49. **Cenchrus,** p. 176
    2. Spikelets not enclosed by a spiny bur.
        3. Spikelets with one or more perfect florets (**Figs. 28 and 29**).
            4. Inflorescence solitary, racemose, paniculate, or spicate, but not digitate.
                5. Inflorescence spicate or spike-like, with one spike per culm_____*Group A*, p. 16
                5. Inflorescence solitary, racemose, or paniculate, but not composed of single spikes.
                    6. Each spikelet with 2 or more perfect florets (**Fig. 29**).
                        7. Some part of the spikelet awned_____
                        _____*Group B*, p. 22
                        7. Spikelet without any awns_____*Group C*, p. 24

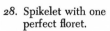

28. Spikelet with one perfect floret.

27. Spiny bur of *Cenchrus.*

29. Spikelet with more than one perfect floret.

6. Each spikelet with one perfect floret (sterile or staminate lemmas may be present, in addition [Fig. 30]).

    8. Some part of the spikelet awned_____
_____Group D, p. 29

    8. Spikelet without any awns_____Group E, p. 34

4. Inflorescence digitate (the spikes and racemes radiating from near the same point [Fig. 31])_____Gorup F, p. 38

3. Spikelets unisexual (i.e., either all staminate or all pistillate)
_____Group G, p. 39

Group A

Inflorescence spicate or spike-like, with one spike per culm; spikelets with one or more perfect florets.

1. Spikelets cylindrical, borne at swollen rachis joints, the entire spikelet falling at maturity; each glume with one awn and one tooth (Fig. 32)_____32. Triticum *

1. Spikelets not as above; rachis joints not swollen; glumes awned or awnless, but not with one awn and one tooth.

    2. Spikelets borne edgewise to the rachis; inner glume absent, except in the terminal spikelet (Fig. 33)_____4. Lolium *

    2. Spikelets borne flatwise to the rachis; glumes present on all spikelets.

        3. Each spikelet subtended and usually surpassed by one or more sterile bristles (not to be confused with awns) (Fig. 34)_____48. Setaria, p. 165

        3. Each spikelet not subtended by bristles.

            4. Each spikelet with two or more perfect florets (Anthoxanthum and Phalaris have three lemmas, but two of them are sterile).

                5. At least some part of the spikelet awned.

                    6. Upper spikelets paired, the lowermost solitary____
_____30. × Agrohordeum *

                    6. Spikelets either all paired, all borne in threes, or all solitary.

                        7. Spikelets either all paired (Fig. 35) or all borne in threes (Fig. 36).

                            8. Spikelets in threes_____29. Hordeum *

                            8. Spikelets paired.

                                9. Glumes to 4 cm long; axis of inflorescence rarely breaking apart at maturity_____
_____27. Elymus *

30. Spikelet with one perfect floret.

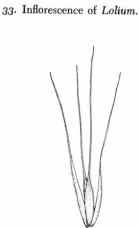

31. Digitate inflorescence.

32. Spikelet of *Triticum*.

33. Inflorescence of *Lolium*.

35. Paired spikelets
of *Elymus*.

34. Spikelet of *Setaria*.

36. Spikelets of
Hordeum.

37. Three-nerved
glume
of *Triticum*.

38. One-nerved
glume of *Secale*.

39. Paired spikele
of *Elymus*.

41. Densely pubescent
lemma of *Tridens*.

40. Rachis with spikelets
in *Agropyron*.

42. Paired
spikelets
of *Elymus*.

9. Glumes 6–8 cm long; axis of inflorescence breaking apart at maturity__28. **Sitanion** *

7. Spikelets solitary.

10. Glumes awned.

11. Glumes 3-nerved (**Fig. 37**)_____ _____32. **Triticum** *

11. Glumes 1-nerved (**Fig. 38**).

12. Awn of lemmas to 8 cm long; annual_____33. **Secale** *

12. Awn of lemmas to 3 cm long; perennials_____31. **Agropyron** *

10. Glumes awnless.

13. Blades 10–20 mm broad; annuals_____ _____32. **Triticum** *

13. Blades 1–10 mm broad; perennials.

14. Awn of lemma more than 1 mm long_____31. **Agropyron** *

14. Awn of lemma up to 1 mm long___ _____9. **Koeleria** *

5. Spikelets awnless throughout.

15. Annuals_____32. **Triticum** *

15. Perennials.

16. Spikelets paired (**Fig. 39**)_____27. **Elymus** *

16. Spikelets solitary.

17. Spikelets borne flatwise to the continuous rachis (**Fig. 40**)_____31. **Agropyron** *

17. Spikelets borne all around the articulated (jointed) rachis.

18. Blades 3–10 mm broad; lemmas densely pubescent on the nerves (**Fig. 41**)_____61. **Tridens**, p. 236

18. Blades 1–3 mm broad; lemmas merely scabrous_____9. **Koeleria** *

4. Each spikelet with a single perfect floret (1–2 sterile lemmas present in addition in *Anthoxanthum* and *Phalaris;* 1 staminate lemma present in addition in *Holcus*).

19. Upper spikelets paired, the lowermost solitary_____ _____30. × **Agrohordeum** *

19. Spikelets either all paired, all borne in threes, or all solitary.

20. Spikelets paired or in groups of three; glumes long-awned.

21. Spikelets paired (**Fig. 42**)____27. **Elymus** *

21. Spikelets in threes (**Fig. 43**)__29. **Hordeum** *

20. Spikelets solitary.

22. Some part of the spikelet awned.

23. Lemmas awnless; glumes awned (**Fig. 44**)_____24. **Phleum** *

23. Lemmas awned; glumes awned or awnless.

24. Lemma awned from the middle (**Fig. 45**)_____16. **Calamagrostis** *

24. Lemma awned from the tip (**Fig. 46**).

25. Lemma 1 per spikelet, perfect.

26. Glumes united near the base (**Fig. 47**)__23. **Alopecurus** *

26. Glumes free at the base_____65. **Muhlenbergia**, p. 244

25. Lemmas 2–3 per spikelet, but only one perfect.

27. Spikelets 5–10 mm long, each with one perfect floret and two empty lemmas (**Fig. 48**)___20. **Anthoxanthum** *

27. Spikelets 3.5–5.0 mm long, each with one perfect floret and one staminate floret (**Fig. 49**)____15. **Holcus** *

22. Spikelets not awned.

28. Glumes 9–15 mm long; lemma 7–14 mm long_____17. **Ammophila** *

28. Glumes to 7 (–10 in *Phalaris*) mm long; lemmas to 7 mm long.

29. Each spikelet with one perfect floret and 1–2 empty lemmas (**Fig. 50**)_____22. **Phalaris** *

29. Each spikelet with one perfect floret only.

30. Lemma 3-nerved (**Fig. 51**)_____65. **Muhlenbergia**, p. 244

30. Lemma 1-nerved.

31. Spikes broad, one-fourth to one-half as broad as long (**Fig. 52**)_____

43. Spikelets of *Hordeum.*

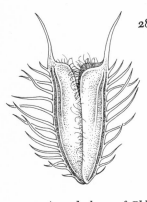

44. Awned glume of *Phleum.*

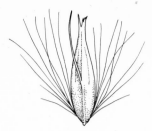

45. Lemma awned from middle in *Calamagrostis*.

47. Glumes united near base in *Alopecurus*.

46. Lemma awned from tip in *Alopecurus*.

48. Spikelet of *Anthoxanthum*.

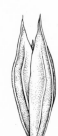

49. Spikelet of *Holcus*.

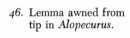

51. Lemma in *Muhlenbergia*.

52. Spike of *Crypsis*.

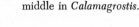

50. Spikelet of *Phalaris*.

53. Spike of *Sporobolus*.

Group B

Inflorescence solitary, racemose, or paniculate, but not spicate or digitate; spikelets with 2 or more perfect flowers; some part of the spikelet awned.

1. Lemmas 2-toothed at the apex (**Fig. 54**).
2. Awn of lemma arising from between the teeth (**Fig. 54**).
3. Lemmas 5- to 9-nerved (**Fig. 55**).
4. Callus of lemmas bearded (**Fig. 56**)____35. **Schizachne** *
4. Callus of lemmas not bearded.
5. Glumes much shorter than the entire spikelet (**Fig. 57**) _____1. **Bromus** *
5. Glumes equalling or longer than the uppermost floret (**Fig. 58**)_____86. **Danthonia**, p. 354

54. Apex of lemma two-toothed.

55. Lemma with an awn between the teeth.

56. Bearded callus of lemma of *Schizachne*.

57. Spikelet with short glumes in *Bromus*.

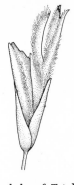

59. Spikelet of *Triplasis*.

60. Spikelet
of *Leptochloa*.

58. Spikelet with long
glumes in *Danthonia*.

    3. Lemmas 3-nerved.
        6. Panicles 3–5 (–8) cm long; spikelets 2- to 5-flowered
          (**Fig. 59**)_____62. **Triplasis,** p. 240
        6. Panicles 10–20 cm long; spikelets 6- to 12-flowered (**Fig.**
          60)_____70. **Leptochloa,** p. 286
  2. Awn of lemma arising near the middle or base of the lemma
    (**Figs. 61 and 62**).

61. Awn arising from
middle of lemma
in *Aira*.

62. Awn arising from base
of lemma in *Deschampsia*.

63. Inflorescence of *Festuca.*

64. Inflorescence of *Dactylis.*

7. Glumes 17–30 mm long; awn of lemmas up to 25 mm long
_____13. **Avena** *
7. Glumes 2.5–5.0 mm long; awn of lemmas 2.5–6.0 mm long.
  8. Awn arising from near the middle of the lemma; lemmas
    3-nerved (**Fig. 61**)_____11. **Aira** *
  8. Awn arising from near base of lemma; lemmas 5-nerved
    (**Fig. 62**)_____12. **Deschampsia** *
1. Lemmas acute or obtuse at the apex, not 2-toothed.
  9. Lemmas 3-nerved_____70. **Leptochloa**, p. 286
  9. Lemmas 5-nerved (all the nerves sometimes obscure in *Festuca*).
    10. Blades involute, about 1 mm in diameter.
      11. Plants annual; stamen 1_____2. **Vulpia** *
      11. Plants perennial; stamens 3_____3. **Festuca** *
    10. Blades flat, 2–8 mm broad.
      12. Lemmas glabrous; spikelets not crowded in 1-sided
        panicles (**Fig. 63**)_____3. **Festuca** *
      12. Lemmas ciliate along the keel; spikelets crowded in
        1-sided panicles (**Fig. 64**)_____8. **Dactylis** *

Group C

Inflorescence solitary, racemose, or paniculate, but not spicate or digitate; spikelets with 2 or more perfect flowers; spikelets awnless.

1. Lemmas distinctly 2-toothed at the apex (**Fig. 65**).
  2. Perennial; blades to 3 mm broad; panicle branches erect or
    spreading; spikelets 5- to 12-flowered; glumes 1 cm long_____
    _____1. **Bromus** *
  2. Annual; blades 5–15 mm broad; panicle branches lax; spikelets

2-flowered; glumes 1.5–2.5 cm long_____13. **Avena** *
1. Lemmas acute to obtuse at the apex, not 2-toothed.
   3. Glumes at least 15 mm long, as long as the spikelets (**Fig. 66**)
    _____13. **Avena** *
   3. Glumes less than 10 mm long, shorter than the spikelets.
      4. Rachilla with long silky hairs, the hairs longer than the spike-
        lets (**Fig. 67**); culms to 4 m tall___84. **Phragmites,** p. 350
      4. Rachilla without long silky hairs longer than the spikelets;
        culms to 1.5 m tall.
        5. Lemmas 3-nerved.
          6. Lemmas 6–10 mm long; grain beaked (**Fig. 68**)____
            _____39. **Diarrhena** *
          6. Lemmas 1.5–5.0 mm long; grain not beaked.
            7. Lemmas glabrous_____60. **Eragrostis,** p. 212
            7. Lemmas pubescent.
              8. Lemmas densely hairy at base, frequently with
                a tuft of hairs.
                9. Lemmas villous at the base, but without a
                  tuft of cobwebby hairs.

68. Beaked grain
of *Diarrhena.*

67. Silky-haired
rachilla of *Phragmites.*

65. Apex of lemma
two-toothed.

66. Spikelet with long
glumes in *Avena.*

          10. Lemmas retuse or obtuse, 3.5–4.0 mm
             long (**Fig. 69**) __6___61. **Tridens,** p. 236
          10. Lemmas acute and mucronate, 4.5 mm
             long (**Fig. 70**)____63. **Redfieldia,** p. 242
         9. Lemmas with a tuft of cobwebby hairs at the
          base, puberulent on the nerves (**Fig. 71**)___
            _____6. **Poa** *

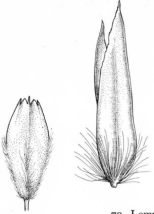

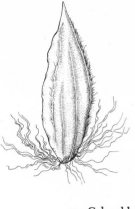

70. Lemmas
of *Redfieldia.*

71. Cobwebby
lemma of *Poa.*

69. Acute and mucronate
lemmas of *Tridens.*

8. Lemmas pubescent only on the nerves.
  11. Lemmas keeled_____60. **Eragrostis**, p. 212
  11. Lemmas rounded on the back.
    12. Lemmas 1.0–2.5 mm long; spikelets 1–5
      mm long_____70. **Leptochloa**, p. 286
    12. Lemmas 4.0 mm long; spikelets 5–8 mm
      long_____61. **Tridens**, p. 236
5. Lemmas 5- to many-nerved, or apparently nerveless, or
  with only the mid-nerve conspicuous.
  13. Lemmas apparently nerveless.
    14. Spikelets disarticulating below the glumes_____
      _____10. **Sphenopholis** *
    14. Spikelets disarticulating above the glumes.
      15. Glumes 2.0–4.5 mm long; plants of moist or
        dry woods_____3. **Festuca** *
      15. Glumes up to 2 mm long; plants of waste
        ground_____5. **Puccinellia** *
  13. Lemmas obviously nerved.
    16. Lemmas 4–10 mm long.
      17. Lemmas as broad as long; inflorescence with
        up to eight spikelets (**Fig. 72**)____7. **Briza** *
      17. Lemmas longer than broad; inflorescence
        with more than eight spikelets.
        18. Lemmas 4–10 mm long, with nine or
          more nerves.

72. Spikelet of *Briza.*

73. Spikelet of
Chasmanthium.

74. Spikelet of *Melica*.

19. Spikelets compressed, 6- to 18-flowered (**Fig. 73**)_____
_____85. **Chasmanthium**, p. 352

19. Spikelets not compressed, 2- to 3-flowered (**Fig. 74**)____34. **Melica** *

18. Lemmas to 7 (–8) mm long, 5- to 7-nerved.

20. Lemmas obscurely 7-nerved (**Fig. 75**); spikelets 10–20 mm long_____
_____35. **Glyceria** *

20. Lemmas 5-nerved; spikelets less than 10 mm long.

21. Spikelets not crowded in 1-sided panicles, not compressed (**Fig. 76**)_____3. **Festuca** *

21. Spikelets crowded in 1-sided panicles, compressed (**Fig. 77**)
_____8. **Dactylis** *

75. Spikelet of *Glyceria*.

76. Inflorescence of *Festuca*.

77. Inflorescence of *Dactylis*.

79. United lodicules of *Glyceria*.

80. Free lodicules of *Puccinellia*.

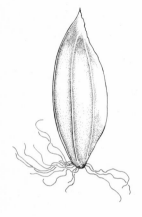

78. Keeled lemma of *Poa*.

81. Inflorescence of *Bouteloua*.

82. Inflorescence of *Aristida*.

83. Spikelet of *Eriochloa*.

16. Lemmas 1.5–5.0 mm long.
    22. Lemmas distinctly keeled (**Fig. 78**)_____
    _____6. **Poa** *
    22. Lemmas rounded on the back.
        23. Nerves of lemma parallel to the summit.
            24. Sheaths closed; lodicules united
                (**Fig. 79**)_____35. **Glyceria** *
            24. Sheaths open; lodicules free from
                each other (**Fig. 80**)_____
                _____5. **Puccinellia** *
        23. Nerves of lemma converging toward the
        summit.
            25. Lemmas glabrous.
                26. Plants annual; stamen 1_____
                    _____2. **Vulpia** *
                26. Plants perennial; stamens 3___
                    _____3. **Festuca** *
            25. Lemmas pubescent, at least on the
                nerves or the keel or at the base__
                _____6. **Poa** *

Group D

Inflorescence solitary, racemose, or paniculate, but not composed of single spikes; each spikelet with one perfect floret (sterile or staminate lemmas may be present, in addition); some part of the spikelet awned.

1. Spikelets borne singly (i.e., not paired).
    2. Lemma 3-awned.
        3. Spikelets borne on one side of a long, arching raceme (**Fig. 81**); lemma rounded on the back; spikelets with one perfect lemma and 1–2 sterile ones_____75. **Bouteloua,** p. 305
        3. Spikelets borne in a more or less erect inflorescence, not 1-sided (**Fig. 82**); lemma inrolled around the palea; no sterile lemma present_____79. **Aristida,** p. 316
    2. Lemma 1-awned or awnless.
        4. First glume reduced to a sheath, united with the lowest, swollen joint of the rachilla (**Fig. 83**)_____44. **Eriochloa** *
        4. First glume not reduced to a sheath and not united with the rachillar joint.
            5. Lemma awnless; glumes awned.
                6. Plants over 1 m tall; lemmas 7–10 mm long_____
                _____77. **Spartina,** p. 312

6. Plants less than 1 m tall; lemmas 2–4 mm long_____
_____65. **Muhlenbergia**, p. 244

5. Lemma awned; glumes awnless or awned.

7. Spikelets arranged in 4 or more crowded ranks, each spikelet composed of one fertile and one sterile floret (**Fig. 84**)_____47. **Echinochloa**, p. 152

7. Spikelets not arranged in 4 or more crowded ranks, each spikelet composed of one fertile floret (also one staminate floret in *Arrhenatherum* or one sterile floret sometimes in *Gymnopogon*).

8. Blades 1–3 mm broad.

9. First glume less than 1 mm long_____
_____65. **Muhlenbergia**, p. 244

9. First glume at least 1.5 mm long.

10. Awn of lemma 2–4 cm long.

11. Tufted annual from a cluster of fibrous roots_____79. **Aristida**, p. 316

11. Cespitose or stout perennial_____
_____37. **Stipa** *

84. Four-ranked spikelets of *Echinochloa.*

10. Awn of lemma up to 2 cm long.

12. Lemma 1.0–1.6 mm long_____
_____18. **Agrostis** *

12. Lemma 2.0–4.5 mm long.

13. Glumes 5-nerved (**Fig. 85**); lemma indurated_____38. **Oryzopsis** *

13. Glumes 1-nerved (**Fig. 86**); lemma not indurated_____
_____65. **Muhlenbergia**, p. 244

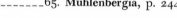

86. Glume of *Muhlenbergia.*

85. Spikelet of *Oryzopsis.*

8. Blades 3 mm broad or broader.
    14. Second glume 5- to 7-nerved.
        15. Awns straight or curved, not twisted near base; second glume 7-nerved (**Fig. 87**)__
                      _____38. **Oryzopsis** *
        15. Awns twisted near base; second glume 5-nerved (**Fig. 88**)_____37. **Stipa** *
    14. Second glume 1- to 3-nerved.
        16. Lemma (excluding awns) 5–10 mm long.
            17. First glume less than 1 mm long_____
                  _____39. **Brachyelytrum** *
            17. First glume 2.5–8.0 mm long.
                18. Awn 10–20 mm long; spikelet (excluding awns) 7–10 mm long; lemma 5- to 7-nerved_____
                      _____14. **Arrhenatherum** *

87. Spikelet of *Oryzopsis.*

                18. Awn to 1.5 mm long; spikelet (excluding awns) 2.5–6.5 mm long; lemma 3-nerved_____19. **Cinna** *
        16. Lemma (excluding awns) 1.5–5.0 mm long.
            19. Spikelets remote along one side of a slender rachis, forming very slender unilateral spikes (**Fig. 89**)_____
                _____71. **Gymnopogon**, p. 296
            19. Spikelets in contracted or open panicles.
                20. Lemma with a tuft of hairs at the base (on the callus), awned from near the middle (**Fig. 90**)_____
                _____16. **Calamagrostis** *

88. Spikelet of *Stipa.*

89. Spike of *Gymnopogon.*

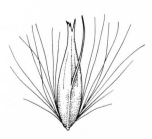

90. Lemma of *Calamagrostis.*

20. Lemma glabrous or pubescent, but without a large tuft of hairs on the callus, awned from the tip.

    21. Plants at least 1 m tall; spikelets disarticulating below the glumes_____19. **Cinna** °

    21. Plants up to 1 m tall, usually smaller; spikelets disarticulating above the glumes_____ ____65. **Muhlenbergia**, p. 244

1. Spikelets borne in pairs.

22. Both spikelets pedicellate, the pedicels unequal in length (**Fig. 91**)_____50. **Miscanthus**, p. 178

22. One spikelet sessile, the other pedicellate (or represented merely by the pedicel).

    23. Pedicellate spikelet represented only by the pedicel (**Fig. 92**)_____53. **Sorghastrum**, p. 195

    23. Pedicellate spikelet present.

        24. Both spikelets of the pair with perfect florets_____ _____51. **Erianthus**, p. 181

        24. Only the sessile spikelet of the pair perfect.

            25. Inflorescence racemose or nearly spicate.

92. Paired spikelets of *Sorghastrum*.

91. Paired spikelets of *Miscanthus*.

93. Inflorescence of *Andropogon*.

95. Inflorescence of *Sorghum*.

94. Inflorescence of *Schizachyrium*.

26. Flowering culms much branched into many short leafy branchlets terminated by 1–6 racemes.
 27. Racemes 2 or more from the sheaths (Fig. 93)_____54. **Andropogon**, p. 196
 27. Raceme solitary at the tip of the peduncle (Fig 94)_____57. **Schizachyrium**, p. 205
26. Flowering culms unbranched_____ _____56. **Bothriochloa**, p. 202
25. Inflorescence paniculate (Fig. 95)_____ _____52. **Sorghum**, p. 186

Group E

Inflorescence solitary, racemose, or paniculate, but not spicate; each spikelet with one perfect floret (sterile or staminate lemmas may be present, in addition); no part of the spikelet awned.

1. Spikelets borne in pairs.
   2. One spikelet of the pair sessile, the other pedicellate (**Fig. 96**)
      _____55. **Microstegium**, p. 202
   2. Both spikelets either sessile or pedicellate.
      3. First glume as long as or longer than the lemmas (**Fig. 97**);
         plants 2.5–4. 0 m tall_____50. **Miscanthus**, p. 178
      3. First glume absent or up to 0.5 mm long, much shorter than
         the lemmas; plants up to 1.5 m tall.
         4. Spikelets with long, tawny hairs longer than the spikelets
            (**Fig. 98**)_____41. **Trichachne** °
         4. Spikelets without long, tawny hairs exceeding the spike-
            lets (**Fig. 99**)_____45. **Paspalum** °
1. Spikelets solitary (i.e., not borne in pairs).
   5. First glume reduced to a sheath and united with the lowest,
      swollen joint of the rachilla (**Fig. 100**)_____44. **Eriochloa** °

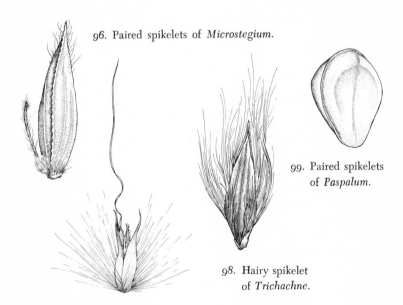

96. Paired spikelets of *Microstegium*.

99. Paired spikelets
    of *Paspalum*.

98. Hairy spikelet
    of *Trichachne*.

97. Spikelet of *Miscanthus*.

5. First glume absent, reduced, or normal, neither sheath-like nor
united with a swollen rachillar joint.
   6. Both glumes absent (**Fig. 101**)_____81. **Leersia**, p. 340
   6. Both glumes present, although the first often much reduced
     or, if absent, the plants not producing seeds.
      7. First glume absent (**Fig. 102**); plants with creeping
        rhizomes, rarely producing seeds_____**Zoysia** [1]
      7. First glume present, although occasionally strongly re-
        duced; rhizome present or absent; plants producing seeds.
        8. First glume up to one-half (to ⅔ in a few species of
          Panicum) as long as second glume.

---

[1] **Zoysia** is frequently planted in Illinois as a choice lawn grass, but no collections have ever been made of it as an adventive. Therefore, it is excluded
from the text.

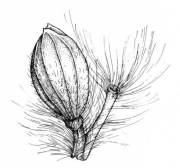

100. Spikelet of *Eriochloa*.

102. Spikelet of *Zoysia*.

101. Spikelet of *Leersia*.

9. Each floret subtended by one or more bristles (**Fig. 103**)_____48. **Setaria**, p. 165

9. Florets not subtended by bristles.

   10. Spikelets solitary at the end of long, capillary pedicels.

      11. Fertile lemma leathery___43. **Leptoloma** °

      11. Fertile lemma indurated_____
      _____46. **Panicum**, p. 43

   10. Spikelets grouped in 2–4 or more ranks.

      12. Inflorescence a dense, dark purple-brown panicle_____47. **Echinochloa**, p. 152

      12. Inflorescence racemose or paniculate with remote, ascending racemes.

         13. Racemes 1–2 (–3) cm long_____
         _____47. **Echinochloa**, p. 152

         13. Racemes over 2 cm long_____
         _____45. **Paspalum** °

8. First glume nearly as long as the second glume, not conspicuously different in size.

   14. Glumes 3-nerved (**Fig. 104**).

      15. Spikelets 4.5–6.0 mm long; glumes 4–6 mm long; blades 2–5 mm broad__21. **Hierochloë** °

      15. Spikelets 2.0–3.5 mm long; glumes 2–3 mm long; some or all the blades over 5 mm broad.

         16. Lemmas 5-nerved; spikelet with one perfect and one sterile lemma (**Fig. 105**); blades to 8 mm broad__26. **Beckmannia** °

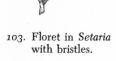

103. Floret in *Setaria* with bristles.

104. Glume of *Hierochloë.*

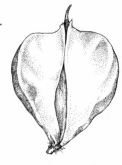

105. Spikelet of *Beckmannia.*

16. Lemma nerveless; spikelet with one perfect lemma (**Fig. 106**); blades to 20 mm broad_____25. **Milium** *

14. Glumes 1-nerved.

17. Lemma with a conspicuous tuft of hairs at the base (on the callus) (**Fig. 107**); spikelets 6–7 mm long_____64. **Calamovilfa**, p. 244

17. Lemma glabrous or pubescent, but without a large tuft of hairs on the callus; spikelets 1–6 mm long.

18. Lemma 3- to 5-nerved, the nerves sometimes obscure.

19. First glume longer than the lemma (**Fig. 108**)_____18. **Agrostis** *

19. First glume shorter than the lemma.

20. Spikelets appressed on two sides of a triangular rachis (**Fig. 109**) _____72. **Schedonnardus**, p. 298

20. Spikelets not confined to two sides of the rachis_____ _____65. **Muhlenbergia**, p. 244

18. Lemma 1-nerved__66. **Sporobolus**, p. 272

106. Spikelet of *Milium.*

107. Lemma of *Calamovilfa.*

108. Spikelet of *Agrostis.*

109. Spikelets of *Schedonnardus.*

Group F

Inflorescence digitate (the spikes and racemes radiating from near the same point).

1. Some part of the spikelet awned.
    2. Spikelets borne in pairs, one sessile and perfect, the other pedicellate and staminate_____54. **Andropogon,** p. 196
    2. Spikelets borne singly.
        3. Spikelets with 3–5 perfect florets; second glume and lemmas awned (**Fig. 110**)_____69. **Dactyloctenium,** p. 286
        3. Spikelets with 1 perfect floret (also 1–2 empty lemmas present); fertile lemma awned (**Fig. 111**)_74. **Chloris,** p. 302
1. Spikelets awnless.
    4. Spikelets with 3–6 perfect florets (**Fig. 112**)_____
    _____68. **Eleusine,** p. 284
    4. Spikelets with 1 perfect floret.
        5. First glume 1.0–1.5 mm long; second glume 1-nerved; no sterile lemmas present_____73. **Cynodon,** p. 300
        5. First glume absent or up to 0.8 mm long; second glume 5-nerved; lower lemma empty_____41. **Digitaria** *

111. Spikelet of *Chloris.*

112. Spikelet of *Eleusine.*

110. Spikelet of *Dactyloctenium.*

Group G

Spikelets unisexual (i.e., either all staminate or all pistillate).

1. Plants to 40 cm tall, dioecious; staminate spikelets 3- to 75-flowered.
    2. Lemmas with a tuft of cobwebby hairs at base (**Fig. 113**)____
    _____6. **Poa** *
    2. Lemmas without a tuft of cobwebby hairs at base.
        3. Both staminate and pistillate spikelets 10- to 75-flowered (**Fig. 114**)_____60. **Eragrostis**, p. 212
        3. Staminate spikelets 3- to 15-flowered, pistillate spikelets 1- to 9-flowered.
            4. Staminate spikelets 3-flowered; pistillate spikelets 1-flowered (**Fig. 115**)_____76. **Buchloë**, p. 310

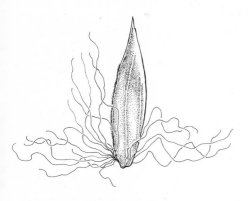

113. Cobwebby lemma in *Poa.*

114. Pistillate and staminate spikelets of *Eragrostis reptans.*

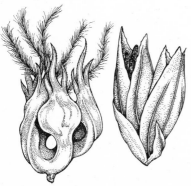

115. Pistillate and staminate spikelets of *Buchloë.*

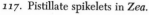

117. Pistillate spikelets in *Zea.*

116. Pistillate and staminate spikelets of *Distichlis.*

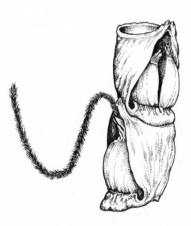

118. Pistillate spikelets in *Tripsacum.*

    4. Staminate spikelets 8- to 15-flowered; pistillate spikelets
        7- to 9-flowered (**Fig. 116**)_____18. **Distichlis,** p. 315
1. Plants 1–4 m tall, monoecious; staminate spikelets 1- to 2-flowered.
    5. Staminate spikelets 2-flowered; glumes membranous.
      6. Annual; staminate and pistillate spikelets in different in-
      florescences; pistillate spikelets borne in pairs (**Fig. 117**)__
      _____59. **Zea,** p. 209
      6. Perennial; staminate and pistillate spikelets in the same in-
      florescence; pistillate spikelets solitary (**Fig. 118**)_____

119. Inflorescence in *Zizania.*

120. Inflorescence in *Zizaniopsis.*

------------------------------58. **Tripsaum,** p. 208
5. Staminate spikelets 1-flowered; glumes none.
   7. Pistillate spikelets confined to the uppermost erect branches
     of the inflorescence, the staminate spikelets confined to the
     lower spreading branches (**Fig. 119**); margin of leaf more
     or less smooth_____82. **Zizania,** p. 345
   7. Pistillate and staminate spikelets on the same branches of the
     inflorescence (**Fig. 120**); margin of leaf harsh and cutting__
     ----------------------------83. **Zizaniopsis,** p. 349

Descriptions and Illustrations

Order Commelinales

POACEÆ – GRASS FAMILY

Annual or perennial herbs (woody in the Bambuseæ); culms cylindrical, with usually hollow internodes and closed nodes; leaves alternate, 2-ranked; sheaths usually free; ligule mostly present; inflorescence composed of (1-) several spikelets; spikelets 1- to several-flowered, each with usually a pair of sterile scales (glumes) at the base; flowers usually perfect, without a true perianth, the perianth reduced to rudiments (lodicules) or absent; flowers subtended by a lemma and a palea; stamens (1-) 3 (-6); ovary 1-celled, with 1 ovule; stigmas 2 (-3); fruit usually a caryopsis (grain).

This family is frequently known as the Gramineæ. It is one of the largest and economically most important families of flowering plants in the world.

In the system of classification followed in this treatment, the Poaceae are one of six families comprising the order Commelinales in Illinois.

Four subfamilies of grasses occur in Illinois.

SUBFAMILY Panicoideae

Annuals or perennials; leaves various; spikelets with one fertile and one sterile or staminate floret, disarticulating below the glumes.

Under the system of classification followed here, the subfamily Panicoideae is composed of tribes Paniceae and Andropogoneae.

Tribe Paniceæ

Annuals, or tufted or rhizomatous perennials; inflorescence a panicle or raceme, sometimes digitate; spikelets with 1 perfect flower; first glume frequently minute or absent; lemma of sterile floret similar in texture to second glume.

42

This tribe is represented in Illinois by nine genera, including *Panicum*, which has the most number of species of any genus of grasses in the state. The other genera are *Digitaria, Trichachne, Leptoloma, Eriochloa*, and *Paspalum*, which have been described in the first (*Bromus to Paspalum*) of these two volumes of grasses, and *Echinochloa, Setaria*, and *Cenchrus*.

46. Panicum L. – Panic Grass

Annuals or perennials; blades flat; inflorescence paniculate; spikelets disarticulating below the glumes; glumes 2, the lower much smaller, the upper more or less equaling and of similar texture to the sterile lemma; sterile lemma with a hyaline palea; fertile lemma indurate, the margins inrolled over the enclosed palea; grain free from the lemma and palea.

Panicum is a huge genus of sometimes poorly differentiated and frequently intergrading taxa. Various treatments of the North American species have been proposed. The most thorough study has been that of Hitchcock and Chase (1910), even though they probably have recognized far too many species. Although their work has been relied upon heavily in this treatment, a somewhat more conservative view is taken in the recognition of species. It is unfortunate, however, that no experimental evidence is available to enable me to make a more reliable placement of the taxa.

The key to taxa which follows is strictly artificial. Unrelated species appear together in the various groups. The technical natural grouping of the species proposed by Hitchcock and Chase is sometimes based on characters difficult to observe and to utilize in a practical key.

Several taxa are extremely rare in Illinois and are known only from one or two localities.

KEY TO THE GROUPS OF Panicum ILLINOIS

1. Spikelets 1.0–1.9 mm long.
    2. Spikelets glabrous_____Group i
    2. Spikelets pubescent_____Group ii
1. Spikelets 2.0 mm long or longer.
    3. Spikelets 2.0–2.9 mm long.
        4. Spikelets glabrous_____Group iii
        4. Spikelets pubescent_____Group iv
    3. Spikelets 3.0 mm long or longer.
        5. Spikelets glabrous_____Group v

5. Spikelets pubescent_____Group vi

Group i

Spikelets 1.0–1.9 mm long, glabrous.
1. Sheaths papillose-hispid.
    2. Fruits stramineous; spikelets 0.9–1.0 mm broad; grain 0.8 mm broad; blades 6–10 mm broad_____3. *P. gattingeri*
    2. Fruits nigrescent; spikelets 0.7 mm broad; grain 0.6 mm broad; blades 2–8 mm broad_____4. *P. philadelphicum*
1. Sheaths glabrous or variously pubescent, but not papillose.
    3. Grain 1.3–1.4 mm long.
        4. First glume up to ¼ as long as the spikelet; nodes of culms with reflexed hairs; spikelets 1.5–1.8 mm long_____
_____17. *P. microcarpon*
        4. First glume nearly ½ as long as the spikelet; nodes of culms glabrous; spikelets at least 1.8 mm long_____8. *P. rigidulum*
    3. Grain 1.8–1.9 mm long.
        5. Culms very slender and wiry; first glume ½ as long as the spikelet; grain 0.7 mm broad; sterile palea enlarging at maturity, and forcing open the spikelet_____12. *P. hians*
        5. Culms not wiry; first glume ¼–⅓ as long as the spikelet; grain 0.8–1.0 mm broad; sterile palea absent, or at least not enlarging at maturity.
            6. Annual; basal and culm leaves similar; basal winter rosette absent; spikelets short-beaked_____1. *P. dichotomiflorum*
            6. Tufted perennial; basal leaves crowded and shorter than culm leaves; spikelets subacute to acute_____
_____20. *P. dichotomum*

Group ii

Spikelets 1.0–1.9 mm long, pubescent.

1. Some or all the sheaths papillose-pilose; spikelets obtuse; blades pubescent throughout on the upper surface (except in some specimens of *P. lanuginosum*).
    2. Culms pilose with horizontally spreading hairs_____
_____25. *P. praecocius*
    2. Culms variously pubescent, but the hairs not horizontally spreading.
        3. Upper surface of blades glabrous except for long hairs at the base; autumnal form matted_____24. *P. lanuginosum*
        3. Upper surface of blades pilose or appressed-pubescent; au-

tumnal form erect or spreading (becoming matted in a
variety of *P. meridionale*).

4. Spikelets 1.3–1.5 mm long; grain 1.2–1.3 mm long.
    5. Panicle branches (at least the lower) drooping, the
       axes long-pilose_____24. *P. lanuginosum*
    5. Panicle branches ascending, the axes glabrous or pu-
       berulent_____23. *P. meridionale*
4. Spikelets 1.6–1.9 mm long; grain 1.5–1.6 mm long.
    6. Upper surface of leaves pilose, the hairs 3–5 mm long;
       first glume acute or acuminate, about ½ as long as the
       spikelet_____26. *P. subvillosum*
    6. Upper surface of leaves short-pubescent, the hairs less
       than 3 mm long; first glume obtuse and truncate,
       ¼–⅓ as long as the spikelet_____24. *P. lanuginosum*
1. None of the sheaths papillose-pilose; spikelets obtuse or acute;
upper surface of blades glabrous except near the base (rarely puberulent throughout in *P. columbianum* and *P. laxiflorum*).
7. Ligule 3–5 mm long_____24. *P. lanuginosum*
7. Ligule less than 2 mm long.
  8. Grain 1.7–1.8 mm long.
    9. Panicle branches ascending, viscid; sheaths viscid; au-
      tumnal culms leafy to the base_____18. *P. nitidum*
    9. Panicle branches spreading, not viscid; sheaths not viscid;
      autumnal culms essentially leafless below the middle____
      _____20. *P. dichotomum*
  8. Grain 1.3–1.5 mm long.
    10. Sheaths retrorsely pilose_____16. *P. laxiflorum*
    10. Sheaths glabrous or ciliate or ascending-pilose.
      11. Sheaths ascending-pilose, with long, soft hairs in-
         termingled with short, crisp ones; blades up to 7 mm
         broad_____29. *P. columbianum*
      11. Sheaths glabrous or ciliate, without two types of
         hairs; blades 7–25 mm broad.
        12. Spikelets ellipsoid, about 0.7 mm broad; nodes
           with reflexed hairs or glabrous.
          13. Spikelets 1.5–1.6 mm long; nodes with re-
             flexed hairs_____17. *P. microcarpon*
          13. Spikelets 1.8 mm long or longer; nodes
             glabrous_____8. *P. rigidulum*
        12. Spikelets obovoid-spherical, 1.0–1.3 mm broad;
           nodes appressed-puberulent or glabrous.
          14. Panicle nearly as broad as long; culms

> spreading; nodes appressed-puberulent___
> _____30. *P. sphaerocarpon*
> 14. Panicle ¼–½ as broad as long; culms erect;
>   nodes more or less glabrous_____
>   _____31. *P. polyanthes*

Group iii

Spikelets 2.0–2.9 mm long, glabrous.

1. Spikelets 1.2–1.7 mm broad.
   2. Blades 3–6 mm broad; spikelets 2.1–2.4 mm long, 1.2–1.3 mm broad_____15. *P. linearifolium*
   2. Blades 6–12 mm broad; spikelets 2.9–3.0 mm long, 1.7 mm broad_____34. *P. oligosanthes*
1. Spikelets 0.7–1.0 mm broad.
   3. At least some of the sheaths papillose-hispid.
      4. Spikelets 2.5–4.0 mm long, acuminate_____5. *P. capillare*
      4. Spikelets up to 2.5 mm long, acute.
         5. Panicle at least half the entire length of the plant_____
            _____5. *P. capillare*
         5. Panicle up to ⅓ the entire length of the plant.
            6. Fruits stramineous; blades 6–10 mm broad; spikelets 0.9–1.0 mm broad; culms stout, to 100 cm tall_____
               _____3. *P. gattingeri*
            6. Fruits nigrescent; blades 2–8 mm broad; spikelets 0.7 mm broad; culms slender, to 50 cm tall_____
               _____4. *P. philadelphicum*
   3. None of the sheaths papillose-hispid.
      7. At least the lower nodes bearded; sheaths softly pubescent
         _____20. *P. dichotomum*
      7. Nodes glabrous or sparsely pilose; sheaths glabrous, ciliate, pilose, or appressed-pubescent, but not softly pubescent.
         8. First glume ⅙–¼ as long as the spikelet_____
            _____1. *P. dichotomiflorum*
         8. First glume ⅓–½ as long as the spikelet.
            9. Ligule ciliate, 2–3 mm long_____10. *P. longifolium*
            9. Ligule 1 mm long or less.
               10. Spikelets 2.5–2.8 mm long; grain short-stipitate__
                  _____9. *P. stipitatum*
               10. Spikelets up to 2.5 mm long (rarely 2.5 mm long in *P. yadkinense*); grain not stipitate.
                  11. Blades pilose on the upper surface near base.
                     12. Blades 1–5 mm broad; palea indurate at

maturity, enlarging and dilating the spike-
let_____12. *P. hians*
12. Blades 5–10 mm broad; palea not indurate
or enlarged at maturity___8. *P. rigidulum*
11. Blades glabrous, except sometimes for marginal cilia.
13. Spikelets 2.2–2.5 mm long.
14. Sheaths with pale glandular spots;
spikelets acute; first glume about ⅓ as
long as spikelet____22. *P. yadkinense*
14. Sheaths without pale glandular spots;
spikelets acuminate; first glume about
½ as long as spikelet__8. *P. rigidulum*
13. Spikelets 2.0–2.2 mm long.
15. First glume about ½ as long as the
spikelet; second glume and sterile
lemma longer than the grain; grain
1.3 mm long, 0.6 mm broad_____
_____8. *P. rigidulum*
15. First glume about ⅓ as long as the
spikelet; second glume and sterile
lemma shorter than the grain; grain
1.8 mm long, 0.9 mm broad_____
_____20. *P. dichotomum*

Group iv

Spikelets 2.0–2.9 mm long, pubescent.

1. Nodes bearded, but with a sticky ring immediately beneath them
_____37. *P. scoparium*
1. Nodes beardless or, if with a beard, then without a sticky ring immediately beneath the nodes.
2. Spikelets up to 2.5 mm long.
3. Ligule 2–5 mm long; sheaths papillose-pubescent.
4. Pubescence of culms horizontally spreading; autumnal
form freely branched; ligule 4–5 mm long_____
_____27. *P. villosissimum*
4. Pubescence of culms appressed or ascending; autumnal
form sparsely branched; ligule 2–3 mm long.
5. Pubescence of blades and sheaths silky; upper surface
of blades pubescent along both margins_____
_____27. *P. villosissimum*
5. Pubescence of blades and sheaths short and stiff; upper

surface of blades more of less glabrous_____
_____28. *P. scoparioides*
3. Ligule 1 mm long or less; sheaths not papillose-pubescent
(except *P. linearifolium*).
  6. Spikelets 1.2–1.5 mm broad (occasionally narrower in *P.*
mattamuskeetense); grain 2.0–2.1 mm long, 1.1–1.2 mm
broad.
    7. Sheaths papillose-pilose; spikelets 1.3–1.5 mm broad
_____15. *P. linearifolium*
    7. Sheaths puberulent, velvety, or glabrous; spikelets
1.2–1.3 mm broad.
      8. Lowermost nodes bearded; lower sheaths velvety-
pubescent_____21. *P. mattamuskeetense*
      8. Nodes without a beard; lower sheaths glabrous or
puberulent, but rarely velvety.
        9. None of the blades over 6 mm broad; spikelets
2.1–2.4 mm long_____15. *P. linearifolium*
        9. Some or all the blades over 6 mm broad; spike-
lets at least 2.4 mm long____38. *P. commutatum*
  6. Spikelets 0.9–1.1 mm broad; grain 1.5–1.9 mm long,
0.9–1.0 mm broad.
    10. Sheaths retrorsely pilose; spikelets papillose-pilose;
grain 1.5 mm long_____16. *P. laxiflorum*
    10. Sheaths puberulent or merely ciliate; spikelets not
papillose; grain 1.7–1.9 mm long.
      11. Upper sheaths viscid-spotted; nodes with reflexed
hairs_____18. *P. nitidum*
      11. Upper sheaths not viscid; nodes glabrous or
sparsely pilose.
        12. Second glume and sterile lemma as long as
the grain; plants green_____19. *P. boreale*
        12. Second glume and sterile lemma shorter than
the grain; plants often purplish_____
_____20. *P. dichotomum*
2. Spikelets 2.5–2.9 mm long.
  13. Sheaths with papillose hairs.
    14. Blades 12–30 mm broad_____40. *P. clandestinum*
    14. Blades 2–12 mm broad.
      15. Blades glabrous or scabrous on the upper surface.
        16. Blades 2–5 mm broad.
          17. Spikelets 1.6–1.7 mm broad; grain 2.4 mm
long, 1.5–1.6 mm broad__14. *P. perlongum*
          17. Spikelets 1.3–1.5 mm broad; grain 2.0–2.1

mm long, 1.2 mm broad_____
_____15. *P. linearifolium*
16. Blades 6–12 mm broad____34. *P. oligosanthes*
15. Blades hirsute or velvety on the upper surface.
18. Blades long-hirsute on both surfaces; grain
2.4–2.5 mm long_____32. *P. wilcoxianum*
18. Blades velvety on both surfaces; grain 2.2 mm
long_____33. *P. malacophyllum*
13. Sheaths without papillose hairs.
19. Lowermost nodes bearded; lower sheaths velvety-
pubescent_____21. *P. mattamuskeetense*
19. None of the nodes bearded; sheaths glabrous or puberulent, but not velvety.
20. Spikelets 2.6–2.8 mm long, obtuse to subacute;
blades firm, more or less cordate at base_____
_____38. *P. commutatum*
20. Spikelets more than 2.8 mm long, abruptly short-
pointed; blades thin, narrowed or slightly rounded
at base_____39. *P. joori*

Group v

Spikelets 3.0 mm long or longer, glabrous.

1. Spikelets obtuse_____34. *P. oligosanthes*
1. Spikelets acute to acuminate.
    2. First glume over ½ as long as the spikelet; grain 2.4–3.0 mm
    long.
        3. Sheaths papillose-hispid; panicle more or less nodding_____
        _____6. *P. miliaceum*
        3. Sheaths ciliate or villous at the throat; panicle ascending or
        spreading_____7. *P. virgatum*
    2. First glume up to ½ as long as the spikelet; grain 1.7–2.3 mm
    long.
        4. Grain 1.4–1.5 mm broad; spikelets 1.5–1.7 mm broad_____
        _____13. *P. depauperatum*
        4. Grain 0.8–1.0 mm broad; spikelets 0.9–1.2 mm broad.
            5. Sheaths glabrous, except for the ciliate margins; first
            glume ⅙–¼ as long as the spikelet___1. *P. dichotomiflorum*
            5. Sheaths papillose-pubescent (occasionally glabrous in *P.
            anceps*); first glume ⅓–½ as long as the spikelet.
                6. Ligule less than 1 mm long; second glume and sterile
                lemma beaked-acuminate; culms glabrous_____
                _____11. *P. anceps*
                6. Ligule 1–3 mm long; second glume and sterile lemma

acuminate, but not conspicuously beaked; culms usually pubescent, at least on the nodes.

  7. Axis of panicle glabrous; grain 2.0 mm long_____
    _____2. *P. flexile*

  7. Axis of panicle short-pilose; grain 1.7–1.8 mm long
    _____5. *P. capillare*

Group vi

Spikelets 3.0 mm long or longer, pubescent.

1. Grain 2.1–2.5 mm long.
  2. Some or all the blades at least 12 mm broad.
    3. Culms and usually the sheaths papillose-hispid; spikelet subacute to acute_____40. *P. clandestinum*
    3. Culms and sheaths glabrous; spikelet abruptly short-pointed
    _____39. *P. joori*
  2. Blades 2–12 mm broad.
    4. Blades glabrous above, glabrous or sparsely pilose beneath.
      5. Blades 6–12 mm broad_____34. *P. oligosanthes*
      5. Blades 2–5 mm broad.
        6. Spikelets acute, the second glume and sterile lemma beaked_____13. *P. depauperatum*
        6. Spikelets obtuse, the second glume and sterile lemma not beaked_____14. *P. perlongum*
    4. Blades long-hirsute or velvety above and beneath.
      7. Blades long-hirsute on both surfaces; grain 2.4–2.5 mm long_____32. *P. wilcoxianum*
      7. Blades velvety on both surfaces; grain 2.2 mm long_____
      _____33. *P. malacophyllum*
1. Grain 2.8–3.5 mm long.
  8. Blades papillose-pubescent on both surfaces; first glume a little more than ½ as long as the spikelet_____36. *P. leibergii*
  8. Blades not papillose-pubescent; first glume up to ½ as long as the spikelet.
    9. Pubescence of sheath not papillose.
      10. Nodes glabrous or sparsely pilose; spikelets 3.0–3.7 mm long_____41. *P. latifolium*
      10. Nodes retrorsely bearded; spikelets 3.8–5.0 mm long___
      _____42. *P. boscii*
    9. Pubescence of sheath papillose.
      11. Nodes not retrorsely bearded_____34. *P. oligosanthes*
      11. Nodes retrorsely bearded.
        12. Ligule 3–4 mm long_____35. *P. ravenelii*

12. Ligule 1 mm long_____42. *P. boscii*

SECTION **Dichotomiflora**

1. **Panicum dichotomiflorum** Michx. Fl. Bor. Am. 1:48. 1803. Annual; culms freely branching, sometimes geniculate at the base, rarely simple, spreading to ascending to erect, to 1.5 m tall, glabrous; sheaths more or less compressed, sometimes inflated, ciliate on the margins toward the summit, otherwise glabrous, very rarely pilose; ligule 1–2 mm long; blades to 20 mm broad, flat or folded, glabrous or sparsely pilose above; panicles terminal and axillary, to 40 cm long, diffuse, the branches ascending, spreading, or reflexed; spikelets 1.7–3.5 mm long, 0.9–1.1 mm broad, oblong-ovoid, short-beaked, usually greenish-purple, glabrous; first glume ⅕–¼ as long as the spikelet, truncate, glabrous; second glume and sterile lemma equal, longer than the grain; grain 1.6–2.0 mm long, 0.8 mm broad, ellipsoid, acute; 2n = 36 (Brown, 1948), 54 (Church, 1929a).

This is the only species of *Panicum* in Illinois which belongs to Section Dichotomiflora, a group characterized by its annual habit and extremely short first glume.

Length of the spikelets is highly variable, ranging from 1.7–3.5 mm. The blades vary from glabrous to sparsely pilose on the upper surface.

Three somewhat intergrading varieties occur in Illinois.

1. Spikelets 2.4–3.5 mm long; at least some of the blades over 5 mm broad.
   2. Culms mostly upright, not geniculate, the nodes not swollen; sheaths not inflated; uppermost panicles long-exserted from the sheaths_____la. *P. dichotomiflorum* var. *dichotomiflorum*
   2. Culms mostly spreading, geniculate, the nodes (at least the lower) swollen; sheaths inflated; uppermost panicles more or less included at base within the sheaths_____ _____lb. *P. dichotomiflorum* var. *geniculatum*
1. Spikelets 1.7–2.3 mm long; blades up to 5 (–8) mm broad_____ _____lc. *P. dichotomiflorum* var. *puritanorum*

1a. Panicum dichotomiflorum Michx. var. **dichotomiflorum** *Fig. 121a–d.*

Culms mostly upright, not geniculate, the nodes not swollen; sheaths not inflated; uppermost panicles long-exserted from the sheaths.

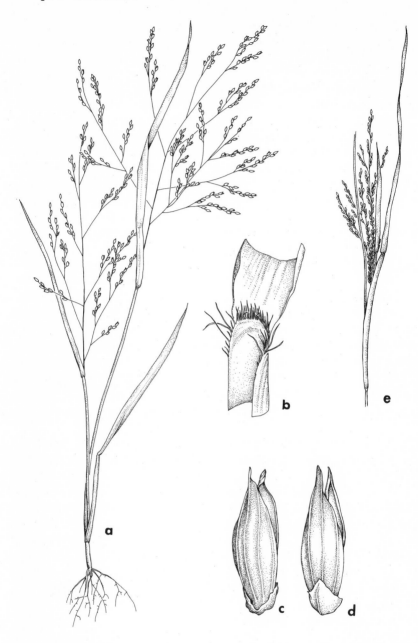

121. Panicum dichotomiflorum (Fall Panicum).—var. *dichotomiflorum. a.*
Habit, X½. *b.* Sheath, with ligule, X5. *c.* Spikelet, front view, X10. *d.*
Spikelet, back view, X10.—var. *geniculatum. e.* Inflorescence, X½.

COMMON NAME: Fall Panicum.

HABITAT: Fields; waste ground, frequently in moist areas.

RANGE: Ontario to Idaho, south to California, Texas, and Florida; West Indies; Mexico.

ILLINOIS DISTRIBUTION: Common throughout the state. This variety is not readily distinguishable from var. *geniculatum*. Some specimens referred here to var. *dichotomiflorum* may be somewhat geniculate, and sometimes the sheaths are a little inflated. The spikelets in var. *dichotomiflorum* are usually less crowded than they are in var. *geniculatum*.

1b. Panicum dichotomiflorum Michx. var. **geniculatum** (Muhl.) Fern. Rhodora 38:387. 1936. *Fig. 121e.*
Panicum geniculatum Muhl. Cat. Pl. 9. 1813.

Culms mostly spreading, geniculate, the nodes (at least the lower) swollen; sheaths inflated; uppermost panicles more or less included at base within the sheaths.

COMMON NAME: Fall Panicum.

HABITAT: Low waste areas and fields.

RANGE: Nova Scotia to Minnesota, south to Louisiana and Florida.

ILLINOIS DISTRIBUTION: Throughout the state, but apparently not as common as var. *dichotomiflorum*.

Although both varieties of *P. dichotomiflorum* may grow together, var. *geniculatum* does not seem to be as common.

The overall range of var. *geniculatum* is generally in the eastern half of the United States, while the range of var. *dichotomiflorum* is throughout the United States.

1c. Panicum dichotomiflorum Michx. var. **puritanorum** Svenson, Rhodora 22:154. 1920. *Fig. 122.*

Culms erect, simple or branching from the base, to 60 cm tall; sheaths glabrous or sparsely pilose; blades to 5 (–8) mm broad; spikelets subacute, 1.7–2.3 mm long.

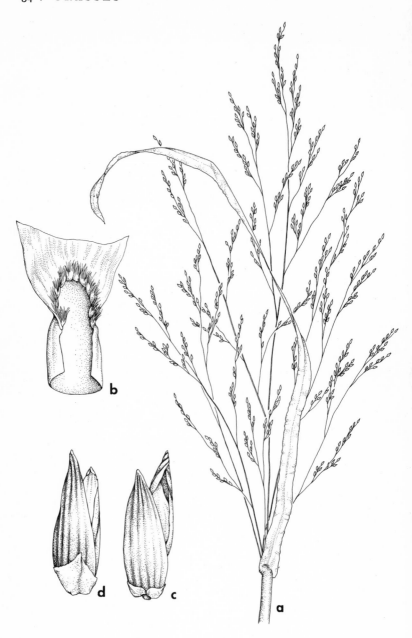

122. *Panicum dichotomiflorum* var. *puritanorum* (Panic Grass). *a.* Inflorescence, X½. *b.* Sheath, with ligule, X2½. *c.* Spikelet, front view, X17½. *d.* Spikelet, back view, X17½.

COMMON NAME: Panic Grass.

HABITAT: Along streams.

RANGE: Nova Scotia to Massachusetts; Indiana; Illinois.

ILLINOIS DISTRIBUTION: Known only from Cook County (*S. Glassman 5591*).

This depauperate variety of *P. dichotomiflorum* resembles *P. gattingeri* by its low, slender habit, its diffuse, ovoid panicles, and its small, long-pedicellate spikelets. It differs from *P. capillare* by its glabrous panicle branches.

SECTION **Capillaria**

2. **Panicum flexile** (Gattinger) Scribn. in Kearney, Bull. Torrey Club 20:476. 1893. *Fig. 123.*

Panicum capillare var. *flexile* Gattinger, Tenn. Fl. 94. 1887.

Annual; culms much branched from the base, erect, slender, to 75 cm tall, glabrous or hispidulous below, the nodes puberulent; sheaths papillose-hispid; ligule 1–2 mm long; blades to 7 mm broad, erect, glabrous or sparsely hispid; panicle to 30 cm long, very narrow, the axis glabrous, the branches mostly ascending; spikelets 3.1–3.5 mm long, 0.9–1.0 mm broad, lanceoloid, acuminate, glabrous; first glume ⅓ as long as the spikelet, acute, glabrous; second glume and sterile lemma subequal, longer than the grain; grain 2.0 mm long, 0.9 mm broad, ellipsoid, subacute; $2n = 18$ (Brown, 1948).

COMMON NAME: Slender Panic Grass.

HABITAT: Moist, sandy soil or calcareous, springy places.

RANGE: Ontario to North Dakota, south to Texas and Florida.

ILLINOIS DISTRIBUTION: Occasional in the southern one-third of the state, rare elsewhere.

This is one of five Illinois species which belong to Section Capillaria, characterized by their annual habit, their glabrous spikelets, and their acute first glumes reaching one-third to one-half the length of the spikelets.

The longer spikelets of *P. flexile* distinguish it from *P. gattingeri, P. philadelphicum,* and *P. capillare* var. *capillare.* It is separated from *P. capillare* var. *occidentale* by its glabrous

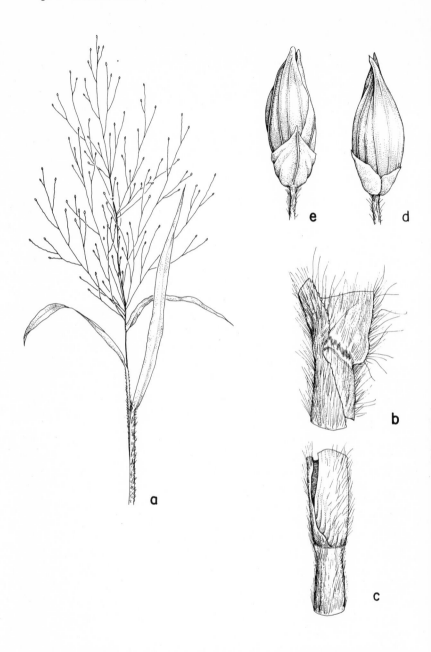

123. *Panicum flexile* (Slender Panic Grass). *a.* Upper part of plant, X½. *b.* Sheath, with ligule, X2½. *c.* Node, X2½. *d.* Spikelet, front view, X10. *e.* Spikelet, back view, X10.

panicle axes. From *P. miliaceum* it differs by its shorter spikelets and erect panicles.

3. **Panicum gattingeri** Nash in Small, Fl. Southeast. U. S. 92. 1903. *Fig. 124.*

> *Panicum capillare* var. *campestre* Gattinger, Tenn. Fl. 94. 1887.
> *Panicum capillare* var. *geniculatum* Scribn. in Kearney, Bull. Torrey Club 20:477. 1893.
> *Panicum capillare gattingeri* Nash in Britt. & Brown, Illustr. Fl. 1:123. 1896.

Annual; culms freely branching, spreading or decumbent, rooting at the lower nodes, to 1 m tall, papillose-hispid; sheaths papillose-hispid; ligule 1–2 mm long; blades 6–10 mm broad, nearly glabrous or hispid on both surfaces; panicles terminal and axillary, to 15 cm long, ⅔ as broad, less than one-half the length of the plant, the branches ascending to spreading; spikelets 1.8–2.5 mm long, 0.9–1.0 mm broad, acute, ellipsoid, glabrous; first glume ⅔ as long as the spikelet, acute or obtuse; second glume and sterile lemma equal, a little longer than the grain; grain stramineous, 1.6 mm long, 0.8 mm broad, ellipsoid, acute.

COMMON NAME: Panic Grass.

HABITAT: Dry or moist open ground, even in waste places.

RANGE: Ontario to Minnesota, south to Arkansas and North Carolina.

ILLINOIS DISTRIBUTION: Scattered throughout the state, except for the northern three tiers of counties.

Gleason (1952) considers this species to be a variety of *P. capillare. Panicum gattingeri* differs from *P. capillare* in its spreading habit, its more numerous, diffuse panicles, and its generally narrower blades; it differs from *P. capillare* var. *occidentale* in its glabrous panicle axes and its slightly longer grains.

Considerable difficulty is encountered in distinguishing *P. gattingeri* from *P. philadelphicum.* In general, however, all the leaves of *P. gattingeri* are at least 6 mm broad, while those of *P. philadelphicum* are less than 6 mm broad.

The spikelets of *P. gattingeri* range from 1.8–2.5 mm in length. The blades may be nearly glabrous or hispid on both surfaces.

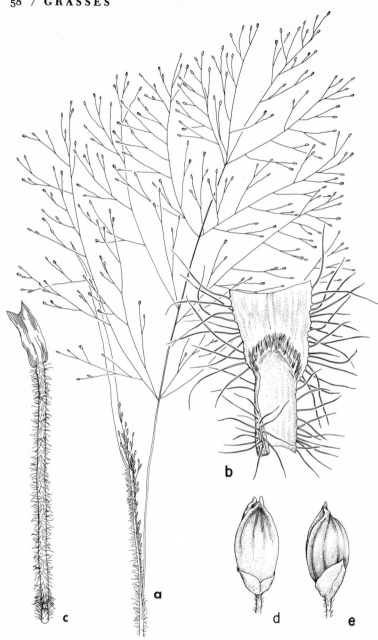

124. Panicum gattingeri (Panic Grass). *a.* Inflorescence, X½. *b.* Sheath, with ligule, X4. *c.* Sheath and node, X1½. *d.* Spikelet, front view, X25. *e.* Spikelet, back view, X12½.

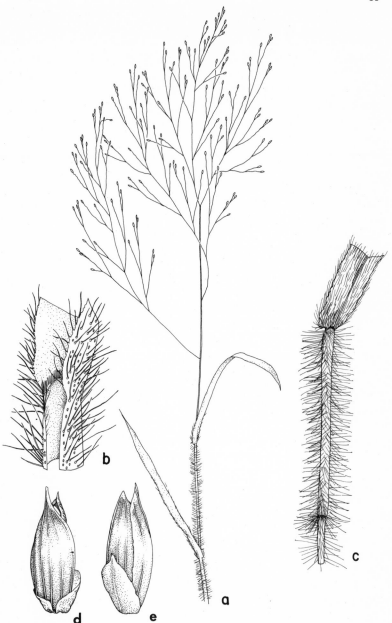

125. Panicum philadelphicum (Panic Grass). *a.* Upper part of plant, X½.
b. Sheath, with ligule, X2½. *c.* Sheath and node, X1½. *d.* Spikelet, front
view, X17½. *e.* Spikelet, back view, X17½.

4. **Panicum philadelphicum** Bernh. ex Trin. Gram. Pan. 216. 1826. *Fig. 125.*
Panicum capillare var. *sylvaticum* Torr. Fl. North. & Mid. U. S. 149. 1824.

Annual in small tufts; culms freely branching, erect to nearly decumbent, to 50 cm tall, papillose-hispid to nearly glabrous; sheaths papillose-hispid; ligule 1–2 mm long; blades 2–8 mm broad, sparsely hirsute to nearly glabrous; panicle generally long-exserted, except in dwarfed specimens, diffuse, to 20 cm long, less than one-half the length of the plant, the branches scabrous and capillary; spikelets 1.7–2.2 mm long, 0.7 mm broad, ellipsoid, acute, glabrous; first glume ⅔ as long as the spikelet, acute; second glume and sterile lemma equal, a little longer than the grain; grain blackish, 1.5 mm long, 0.6 mm broad, ellipsoid, acute; 2n = 18 (Brown, 1948).

COMMON NAME: Panic Grass.
HABITAT: Dry, usually sandy, soil.
RANGE: Connecticut to Minnesota, south to Texas and Georgia.
ILLINOIS DISTRIBUTION: Locally scattered in the southern three-fifths of the state; also DeKalb County.
This species differs from *P. flexile* in its shorter spikelets, from *P. gattingeri* in its narrower blades and more slender culms, from *P. capillare* var. *capillare* in its smaller and more spreading panicles, and from *P. capillare* var. *occidentale* in its glabrous or scabrous panicle axes. The greatest confusion on identifications is between dwarfed specimens of *P. philadelphicum*, which have short-exserted panicles, and *P. gattingeri*. These individuals are best distinguished on the basis of leaf widths.

5. **Panicum capillare** L. Sp. Pl. 58. 1753.

Annual; culms sparsely or freely branched, erect to ascending, to 75 cm tall, papillose-hispid to nearly glabrous, the nodes densely pubescent; sheaths densely papillose-hispid; ligule 1–2 mm long; blades to 15 mm broad, hispid to nearly glabrous above and below; panicle ½–⅔ the length of the plant, diffuse, densely flowered, the capillary branches ascending to spreading, the axis hispid, often becoming reddish-purple at maturity; spikelets 2–4 mm long, 0.8–1.0 mm broad, short-acuminate, lance-ellipsoid to ellipsoid, glabrous; first glume about ⅓–½ as long as the spikelet, acute, glabrous or scabrous on the midnerve at the apex; second glume and sterile lemma subequal to equal, extending beyond the grain; grain stramineous, 1.5–1.8 (–2.1) mm long, 0.7–0.9 mm broad, ellipsoid, obtuse to acute.

This species is distinguished by its short-acuminate spikelets and its panicles which are as broad as or broader than long.

In this work I am considering *P. capillare* and *P. barbipulvinatum* as varieties of the same species, in line with most recent opinions of grass experts. These two varieties are distinguished as follows:

1. Lowest branches of panicle included at the base; spikelets mostly 2.0–2.5 mm long; grain about 1.5 mm long_____
   _____5a. *P. capillare* var. *capillare*
1. Panicle long-exserted; spikelets mostly 2.5–4.0 mm long; grain 1.6–1.7 mm long_____5b. *P. capillare* var. *occidentale*

5a. Panicum capillare L. var. **capillare** *Fig. 126.*

Culms sparsely branched, erect or ascending, to 75 cm tall, papillose-hispid to nearly glabrous, the nodes densely pubescent; blades to 15 mm broad, hispid above and below; panicle ½–⅔ the length of the plant, diffuse, densely flowered, the capillary branches ascending to spreading; spikelets 2.0–2.5 mm long, 0.8–0.9 mm broad, ellipsoid; first glume about ½ as long as the spikelet, glabrous; second glume and sterile lemma equal; grain 1.5 mm long, 0.7–0.8 mm broad, acute; 2n = 18 (Avdulov, 1928).

COMMON NAME: Witch Grass.

HABITAT: Fields, waste ground.

RANGE: Quebec to Montana, south to Texas and Florida; Bermuda.

ILLINOIS DISTRIBUTION: Common throughout the state. This variety tends to intergrade into var. *occidentale* on characters of the panicle and the spikelets so that it is not feasible to maintain both as separate species.

5b. Panicum capillare L. var. **occidentale** Rydb. Contr. U. S. Natl. Herb. 3:186. 1895. *Fig. 127.*

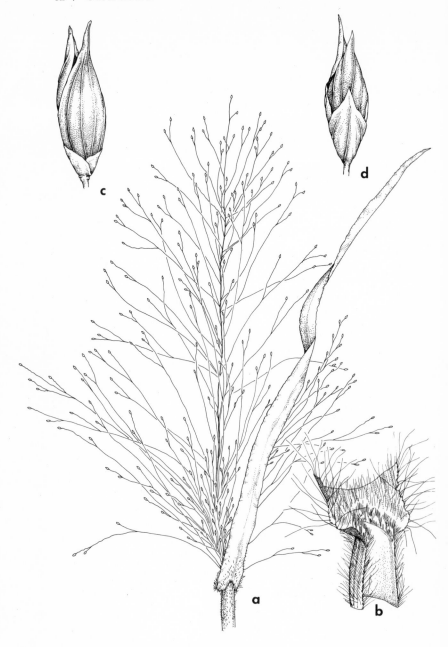

126. Panicum capillare var. *capillare* (Witch Grass). *a.* Inflorescence,
X½. *b.* Sheath, with ligule, X2½. *c.* Spikelet, front view, X15. *d.* Spikelet,
back view, X15.

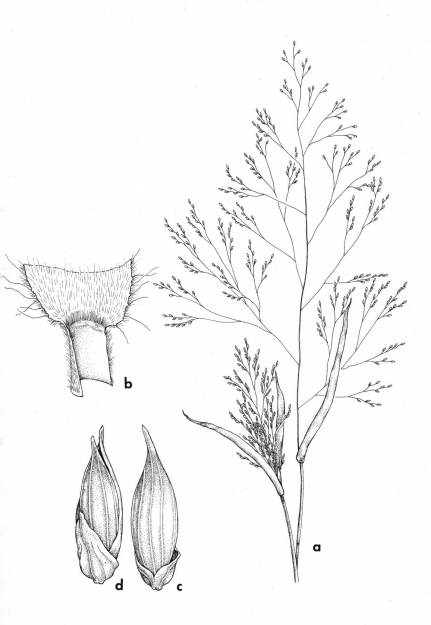

127. *Panicum capillare* var. *occidentale* (Witch Grass). *a.* Inflorescence, X½. *b.* Sheath, with ligule, X2½. *c.* Spikelet, front view, X15. *d.* Spikelet, back view, X15.

Panicum capillare brevifolium Vasey ex Rydb. & Shear, Bull.
U.S.D.A. Div. Agrost. 5:21. 1897.
Panicum barbipulvinatum Nash in Rydb. Mem. N. Y. Bot.
Gard. 1:21. 1900.

Culms freely branching from near the base, erect, to 50 cm tall,
glabrous or hispid below the nodes; blades to 12 mm broad,
hispid or nearly glabrous on both surfaces; panicle about ½ the
length of the plant, the branches ascending; spikelets 2.5–4.0 mm
long, 1 mm broad, lance-ellipsoid; first glume ⅔ as long as the
spikelet, scabrous on the midnerve at the apex; second glume and
sterile lemma subequal; grain 1.7–1.8 (–2.1) mm long, 0.9 mm
broad, obtuse to subacute.

COMMON NAME: Witch Grass.
HABITAT: Open ground.
RANGE: Wisconsin to British Columbia, south to California and Texas.
ILLINOIS DISTRIBUTION: Very rare; known from Henderson County (Oquawka, *H. N. Patterson s.n.*).
The spikelets and grains are slightly larger than in *P.
capillare* var. *capillare*, and some intergradation does
occur.
Extreme specimens of var. *occidentale*, with spikelets
4 mm long and with the panicles long-exserted, look quite distinct from var. *capillare*.

6. Panicum miliaceum L. Sp. Pl. 58. 1753. *Fig. 128.*

Annual; culms erect or decumbent, branching from the lower
nodes, to 1 m tall, hispid or glabrous except for the puberulent
nodes; sheaths papillose-hispid; ligule 1–2 mm long; blades to
20 mm broad, pilose or glabrous on both surfaces; panicle to 30
cm long, more or less nodding, the branches narrowly ascending;
spikelets 4.5–5.5 mm long, 2.0–2.7 mm broad, ovoid, acuminate,
glabrous; first glume at least ½ as long as the spikelet, acuminate,
glabrous; second glume and sterile lemma subequal, longer than
the grain; grain about 3 mm long, about 2 mm broad, ellipsoid,
subacute; 2n = 36 (Avdulov, 1931).

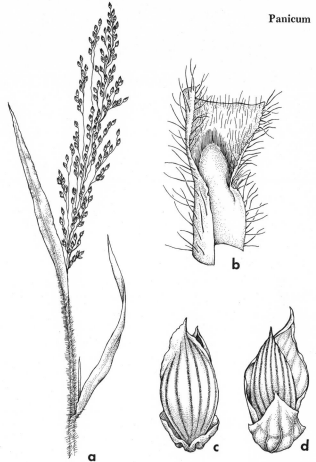

128. *Panicum miliaceum* (Broomcorn Millet). *a.* Upper part of plant, X½. *b.* Sheath, with ligule, X2½. *c.* Spikelet, front view, X10. *d.* Spikelet, back view, X10.

COMMON NAME: Broomcorn Millet.

HABITAT: Waste ground, usually escaped from cultivation.

RANGE: Native of Europe and Asia; occasionally established in the United States.

ILLINOIS DISTRIBUTION: Not common; scattered in the northern and central counties; also Jackson County.

This species has the largest spikelets in the genus in Illinois, with an average length of 5 mm. Variation exists in the size of the plant. Specimens as short as 10 cm and

129. *Panicum virgatum* (Switch Grass). *a*. Inflorescence, X½. *b*. Sheath, with ligule, X2½. *c*. Spikelet, front view, X12½. *d*. Spikelet, back view, X12½.

as tall as one meter have been found with mature grains in Illinois. Pubescence of the culms and blades of the leaves is variable.

This species was introduced into this country from the Old World as a forage crop, although it is no longer extensively planted in Illinois. The first Illinois collection apparently was made just before the end of the nineteenth century.

Broomcorn Millet flowers from June to late October.

SECTION **Virgata**

7. **Panicum virgatum** L. Sp. Pl. 59. 1753. *Fig. 129.*

Mostly tufted perennial from scaly rhizomes; culms erect, to nearly 2 m tall, glabrous, often glaucous; sheaths ciliate to villous at the throat, otherwise glabrous; ligule 2–4 mm long; blades to 15 mm broad, glabrous or pilose near base above, rarely pilose throughout, glabrous below, the margins scabrous; panicle to 50 cm long, ⅓–½ as broad, the branches ascending to spreading; spikelets 3.5–6.0 mm long, 1.2–1.5 mm broad, ellipsoid-ovoid, acuminate, glabrous; first glume about ⅔ as long as the spikelet, acuminate to cuspidate, glabrous; second glume and sterile lemma very unequal, the sterile lemma longer than the grain; grain 2–3 mm long, 1.0–1.5 mm broad, narrowly ovoid to ellipsoid, subacute; 2n = 21, 25, 30, 32 (Brown, 1948), 36, 72 (Church, 1929).

COMMON NAME: Switch Grass.

HABITAT: Fields, waste ground, prairies, moist seepage areas of cliffs, rocky stream beds, woods.

RANGE: Nova Scotia to North Dakota and Nevada, south to Arizona, Texas, and Florida; Mexico; Central America.

ILLINOIS DISTRIBUTION: Rather common throughout the state.

A few specimens have the upper surface of the blades pilose throughout; some specimens appear to be glaucous, while others do not. This species sometimes attains a height in Illinois of nearly two meters, thereby making it one of the most robust species of *Panicum* in the state.

As the only representative of Section Virgata in Illinois, *P. virgatum* is the only Illinois *Panicum* with a perennial habit, long-pedicelled spikelets, and a diffuse panicle.

Panicum virgatum shows a great tolerance for a wide variety of environmental conditions. Specimens growing in moist depressions of cliffs are frequently dwarfed and bear generally unexpanded panicles.

SECTION **Agrostoidea**

8. **Panicum rigidulum** Bosc ex Nees in Mart. Fl. Bras. 2 (1):163. 1829.

Panicum agrostoides Spreng. Pl. Pugill. 2:4. 1815, nomen illeg. Densely clumped perennial from a short caudex; culms erect, compressed, often geniculate at base, glabrous, to 100 cm tall; sheaths pilose on the sides, or appressed-pubescent, or glabrous; ligule erose, about 1 mm long; blades flat above, to 12 mm broad, becoming folded near the base, glabrous or sparsely pilose above, near the base; panicles terminal and axillary, to 30 cm long, ½–⅔ as broad, the axis glabrous to scabrous, the branches spreading, ascending, or strongly erect; spikelets 1.8–2.5 mm long, 0.7–0.8 mm broad, narrowly ellipsoid to lanceoloid, acute to short-acuminate, glabrous or with the nerves somewhat scabrous; first glume about ½ as long as the spikelet, acute or acuminate, glabrous; second glume and sterile lemma subequal, with scabrous nerves near the tips; grain 1.3–1.5 mm long, 0.6–0.7 mm broad, ellipsoid, apiculate.

This is the species which is known throughout its range as *P. agrostoides* Spreng. Voss (1966) explains in considerable detail why this latter binomial must be rejected in favor of *P. rigidulum.*

Panicum rigidulum is distinguished by its compressed culms, its spikelets mostly arranged on one side of the axes of the inflorescence (secund), and by its shiny spikelets.

Two varieties may be distinguished in Illinois, separated as follows:

1. Panicle branches spreading to ascending; spikelets 1.8–2.2 mm long_____8a. *P. rigidulum* var. *rigidulum*
1. Panicle branches strongly erect; spikelets 2.2–2.5 mm long_____ _____8b. *P. rigidulum* var. *condensum*

8a. **Panicum rigidulum** Nees var. **rigidulum** *Fig. 130.*

Panicum elongatum Pursh var. *ramosior* Mohr, Contr. U. S. Nat. Herb. 6:357. 1901.
Panicum agrostoides Spreng. var. *ramosius* (Mohr) Fern. Rhodora 38:390. 1936.

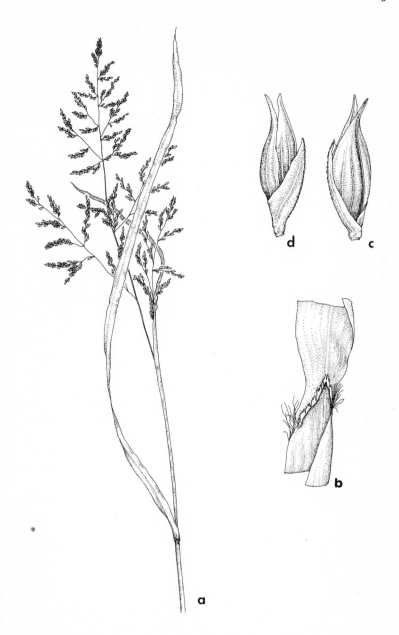

130. Panicum rigidulum var. *rigidulum* (Munro Grass). *a.* Upper part of plant, X½. *b.* Sheath, with ligule, X2½. *c.* Spikelet, front view, X17½. *d.* Spikelet, back view, X17½.

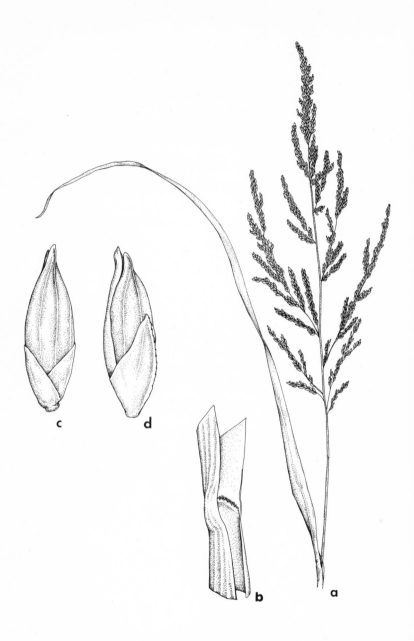

131. Panicum rigidulum var. *condensum* (Panic Grass). *a.* Inflorescence, X½. *b.* Sheath, with ligule, X2½. *c.* Spikelet, front view, X17½. *d.* Spikelet, back view, X17½.

Culms to 1 m tall; sheaths pilose on the sides, otherwise glabrous; blades to 12 mm broad, glabrous; panicle to 30 cm long, the axis often glabrous, spreading to ascending; spikelets 1.8–2.2 mm long, 0.7–0.8 mm broad, narrowly ellipsoid; grain 1.3 mm long, 0.6 mm broad; 2n = 18 (Brown, 1948).

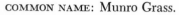

COMMON NAME: Munro Grass.

HABITAT: Moist soil in prairies, along ponds or creeks, or even in low woodlands.

RANGE: Maine to Kansas, south to Texas and Florida; British Honduras.

ILLINOIS DISTRIBUTION: Occasional in the southern one-third of the state; not common in the central one-third of the state; rare in the northern one-third of the state. Nearly all Illinois specimens have spikelets 1.8–2.0 mm long, although a few specimens from Illinois were seen with spikelets up to 2.2 mm long.

Panicum rigidulum var. *rigidulum* differs from var. *condensum* in its more spreading panicle branches and in its smaller spikelets. Some authors, such as Hitchcock (1950), consider *P. condensum* to be a distinct species, while others do not even give it varietal status.

Panicum agrostoides var. *ramosius*, attributed to southern Illinois by Fernald (1950), reputedly differs from var. *rigidulum* by its darker nodes and its greenish or lead-colored long-tipped spikelets. I have been unable to correlate these differences in Illinois specimens; therefore var. *ramosius* is included under var. *rigidulum*.

8b. Panicum rigidulum Nees var. **condensum** (Nash) Mohlenbrock, comb. nov. *Fig. 131.*

Panicum condensum Nash in Small, Fl. Southeast. U. S. 93. 1903.

Panicum agrostoides Spreng. var. *condensum* (Nash) Fern. Rhodora 36:74. 1934.

Culms to 75 cm tall; sheaths appressed-pubescent or glabrous; blades to 10 mm broad, sparsely pilose at the base of the upper surface, glabrous below; panicle to 25 cm long, very narrow, the axis scabrous, the branches strongly erect; spikelets 2.2–2.5 mm long, 0.8 mm broad, lanceloid, acuminate, glabrous except for the scabrous nerves; grain 1.4–1.5 mm long, 0.7 mm broad; 2n = 18 (Brown, 1948).

COMMON NAME: Panic Grass.

HABITAT: Moist soil.

RANGE: Massachusetts to Kansas, south to Texas and Florida.

ILLINOIS DISTRIBUTION: Rare; first collected from Massac County (Mermet Conservation Lake, June 27, 1964, *R. H. Mohlenbrock 12626*). This taxon is common in Johnson County along the Cache River near Heron Pond where it was first found by J. White in 1969.

Most workers consider this taxon to be a variety of *P. rigidulum,* although Hitchcock (1950) continues to maintain it as a species. It has a much narrower and more erect panicle and longer spikelets than var. *rigidulum.*

9. **Panicum stipitatum** Nash in Scribn. Bull. U.S.D.A. Div. Agrost. 17:56. 1901. *Fig. 132.*

Panicum elongatum Pursh, Fl. Am. Sept. 1:69. 1814, non Salisb. (1796).

Panicum agrostoides Spreng. var. *elongatum* (Pursh) Scribn. Tenn. Agr. Exp. Sta. Bull. 7:42. 1894.

Densely tufted perennial from a short caudex; culms compressed, erect, to nearly 1 m tall, glabrous, often purplish; sheaths glabrous or pilose on the sides; ligule about 1 mm long; blades to 12 mm broad, scabrous on the lower surface; panicle to 20 cm long, ⅓–½ as wide, often purple, the branches ascending; spikelets 2.5–2.8 mm long, 0.7 mm broad, lance-ellipsoid, acute and curved at the tip, glabrous; first glume about ½ as long as the spikelet, acute, glabrous; second glume and sterile lemma subequal, much longer than the grain; grain 1.5 mm long, 0.6 mm broad, ellipsoid, apiculate, short-stipitate; 2n = 18 (Brown, 1948).

COMMON NAME: Panic Grass.

HABITAT: Moist soil.

RANGE: Connecticut to Missouri, south to Texas and Florida.

ILLINOIS DISTRIBUTION: Rare; known from Johnson County (low ground, one mile west of West Vienna, June 27, 1964, *R. H. Mohlenbrock 12634*).

This species is related to *P. rigidulum* from which it differs by its more diffuse panicles and its short-stipitate grains.

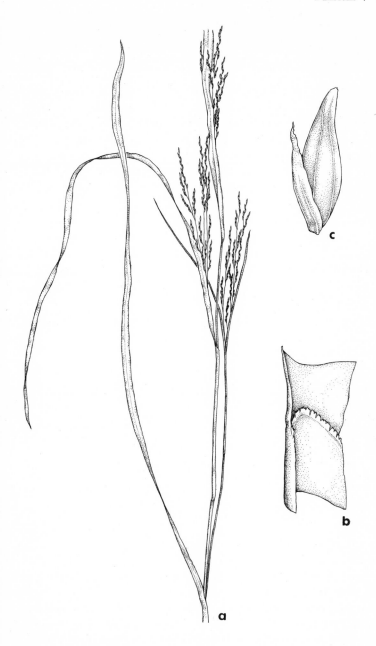

132. Panicum stipitatum (Panic Grass). *a.* Upper part of plant, X½. *b.* Sheath, with ligule, X4. *c.* Spikelet, X20.

It is extremely rare in the western part of its range. Gleason (1952) calls this taxon *P. agrostoides* var. *elongatum*.

10. **Panicum longifolium** Torr. Fl. North. & Mid. U. S. 149. 1824. *Fig. 133.*

Densely tufted perennial; culms slender, compressed, stiff, erect, to 85 cm tall, glabrous; sheaths keeled, pubescent at the junction with the blade, glabrous or villous otherwise; ligule 2–3 mm long; blades 2–5 mm broad, erect or recurved, pilose near the base above, glabrous or pilose below; panicle to 25 cm long, ½–⅔ as wide, the branches scabrous and ascending; spikelets 2.0–2.7 (–3.0) mm long, 0.7 mm broad, lance-ellipsoid, acute to short-acuminate, glabrous to scabrous at the tip; first glume ⅖–½ as long as the spikelet, acute; second glume and sterile lemma subequal, keeled, longer than the grain; grain 1.6–1.7 mm long, 0.6 mm broad, ellipsoid, obtuse; 2n = 18 (Brown, 1948).

COMMON NAME: Panic Grass.

HABITAT: Rocky ledges in a wooded ravine on top of a limestone cliff.

RANGE: Massachusetts to Ohio to Illinois, south to Texas and Florida.

ILLINOIS DISTRIBUTION: Very rare; known only from Monroe County (Fults, October 25, 1962, *J. Ozment, R. H. Mohlenbrock, & W. Crews 12799*).

The long-ciliate ligule separates this species from *P. rigidulum* and other related species of Section Agrostoidea. Some of the spikelets of the Illinois specimen measure up to 3 mm long, and some of the panicle branches tend to spread a little. Both of these characters are found in var. *combsii*, which our plant strongly resembles, but which ranges to the southeast of Illinois.

11. **Panicum anceps** Michx. Fl. Bor. Am. 1:48. 1803. *Fig. 134.*

Tufted perennial from stout, scaly rootstocks; culms erect, to 1 m tall, glabrous; sheaths glabrous to densely papillose-pilose; ligule less than 1 mm long; blades to 12 mm broad, glabrous, scabrous, or pilose; panicle to 40 cm long, ⅓–⅔ as broad, the branches spreading to ascending; spikelets 3.0–3.8 mm long, 1.0–1.2 mm broad, lance-ellipsoid, curved at the short-acuminate tip, glabrous; first glume ⅓–½ as long as the spikelet, acute, glabrous; second glume and sterile lemma subequal, longer than the grain;

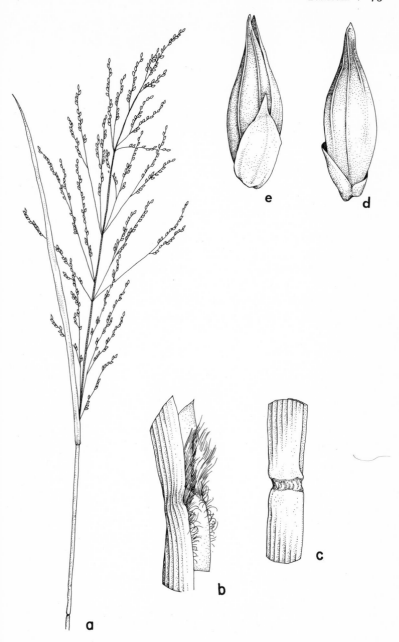

133. *Panicum longifolium* (Panic Grass). *a.* Inflorescence, X½. *b.* Sheath, with ligule, X5. *c.* Sheath and node, X5. *d.* Spikelet, front view, X17½. *e.* Spikelet, back view, X17½.

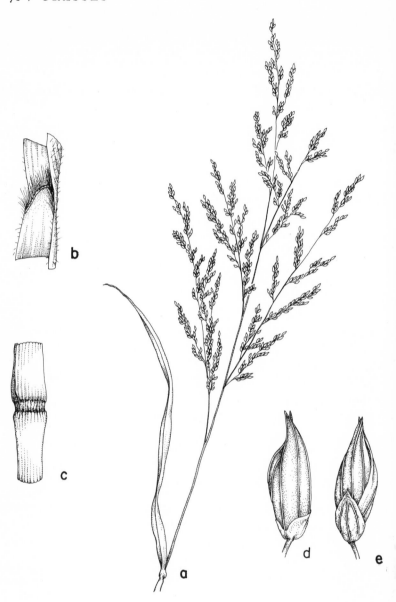

134. Panicum anceps (Panic Grass). *a.* Inflorescence, X¼. *b.* Sheath, with ligule, X5. *c.* Sheath and node, X5. *d.* Spikelet, front view, X10. *e.* Spikelet, back view, X10.

grain 2.0–2.2 mm long, 1.0 mm broad, ellipsoid, subacute; 2n = 18 (Brown, 1948).

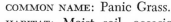

COMMON NAME: Panic Grass.

HABITAT: Moist soil, occasionally in woodlands; wet prairies, roadside ditches, banks of streams.

RANGE: New Jersey to Kansas, south to Texas and Florida.

ILLINOIS DISTRIBUTION: Common in the southern one-third of the state, rare to absent elsewhere.

Variation exists in the degree of pubescence of the sheaths and blades.

This species is one of the easiest Panicums to identify because of the large, curved spikelets. It often grows with *P. rigidulum,* a similar species with straight spikelets less than 3 mm long.

SECTION **Laxa**

12. Panicum hians Ell. Bot. S.C. & Ga. 1:118. 1816. *Fig. 135.*

Cespitose perennial; culms sparsely branched, erect or geniculate at the base, to 60 cm tall; sheaths glabrous, keeled; ligules 0.5 mm long; blades 1–5 mm broad, flat or occasionally folded, pilose on the upper surface near the base, glabrous below; panicle to 20 cm long, loose and open, the branches ascending, spreading, or drooping; spikelets 1.5–2.4 mm long, 0.8 mm broad, glabrous, more or less secund; first glume about ½ as long as the spikelet, acute; second glume and sterile lemma subequal; palea indurate and enlarged at maturity, forcing the spikelet open; grain 1.4–1.9 mm long, 0.7 mm broad, ellipsoid, acute; 2n = 18 (Brown, 1948).

COMMON NAME: Panic Grass.

HABITAT: Low roadside ditch.

RANGE: Virginia to southern Illinois to Oklahoma, south to New Mexico and Florida; Mexico.

ILLINOIS DISTRIBUTION: Rare; known only from a low roadside ditch in Alexander County (vicinity of Gale, June 30, 1968, *R. H. Mohlenbrock 13004*).

This is the only species of *Panicum* in Illinois in which the palea becomes enlarged and indurate, expanding the spikelet at maturity. It is the sole member of Section Laxa.

At its only station in Illinois, *P. hians* was found growing with

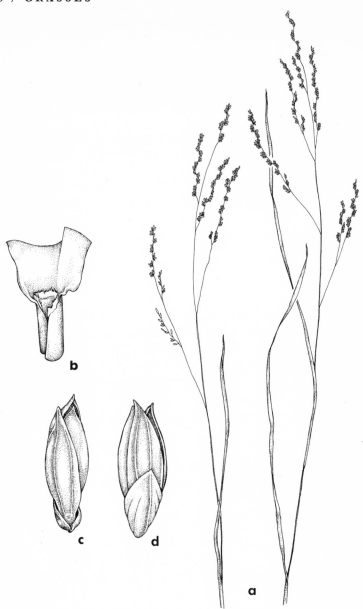

135. Panicum hians (Panic Grass). *a.* Upper part of plants, X½. *b.* Sheath, with ligule, X4. *c.* Spikelet, front view, X17½. *d.* Spikelet back view, 17½.

Eleocharis palustris, Ammannia coccinea, and *Ludwigia palustris* var. *americana.* In the southeastern United States, where it is more abundant, it is a frequent inhabitant of cypress swamps.

SECTION **Depauperata**

13. **Panicum depauperatum** Muhl. Descr. Gram. 112. 1817. *Fig. 136.*

Tufted perennial; culms erect or spreading, to 40 cm tall, glabrous, puberulent, or pilose, the nodes ascending-pubescent; sheaths glabrous to papillose-hirsute; ligule about 0.5 mm long or less; blades to 5 mm broad, occasionally pubescent below, otherwise glabrous, frequently drying involute; panicle to 10 cm long, very narrow, few-flowered, the branches strict and ascending; spikelets 3–4 mm long, 1.5–1.7 mm broad, ellipsoid, acutely beaked, glabrous or sparsely pubescent; first glume ⅓–½ as long as the spikelet, subacute, glabrous or sparsely pubescent; second glume and sterile lemma equal, forming more or less a beak, longer than the grain; grain 2.1–2.3 mm long, 1.4–1.5 mm broad, ovoid, umbonate at the apex; autumnal form similar to the vernal form, the panicles reduced; 2n = 18 (Brown, 1948).

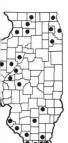

COMMON NAME: Panic Grass.

HABITAT: Dry, open woodlands; prairies, waste ground, sandy soil in black oak woods.

RANGE: Nova Scotia to Minnesota, south to Texas and Georgia.

ILLINOIS DISTRIBUTION: Occasional throughout the state. Section Depauperata is characterized by species with blades more than twenty times longer than broad and not exceeding 5 mm in width. Of the three species of this section in Illinois, *P. depauperatum* is the only one with an acutely beaked spikelet. In addition, the second glume and sterile lemma are much longer than the grain.

Panicum depauperatum shows considerable variability in the size and pubescence (or lack of it) of the spikelets. The vernal and autumnal phases are rather different. In the vernal stage, the culms are mostly simple and the panicles exserted with strongly ascending branches, while in the autumnal stage, the culms fork near the base and the panicles are much reduced and confined to the lower axils and the basal sheaths.

In dry woodlands, where *P. depauperatum* seems to occur more commonly, the associated species include *Cunila origano-*

ides, Gillenia stipulata, Antennaria plantaginifolia, Verbesina alternifolia, Muhlenbergia sobolifera, and *Carex retroflexa.*

Vernal forms may be found in flower as early as mid-May, while autumnal phases begin to flower during July.

14. Panicum perlongum Nash, Bull. Torrey Club 26:575. 1899. *Fig. 137.*

Tufted perennial; culms erect or spreading, to 40 cm tall, pilose; sheaths papillose-pubescent; ligule about 0.5 mm long; blades 2–5 mm broad, pilose below; panicle to 10 cm long, narrow, the branches erect; spikelets 2.7–3.2 mm long, 1.6–1.7 mm broad, ovoid, obtuse, sparsely pilose; first glume ¼–⅓ as long as the spikelet, acute or obtuse; second glume and sterile lemma equal, as long as the grain; grain 2.4 mm long, 1.5–1.6 mm broad, obovoid to ovoid, umbonate at the apex; autumnal form similar to the vernal form, but with more panicles.

COMMON NAME: Panic Grass.

HABITAT: Dry soil, particularly in prairies or upland woods.

RANGE: Michigan to Manitoba, south to Colorado and Texas and Indiana.

ILLINOIS DISTRIBUTION: Occasional in the northern half of the state, rare in the southern half.

This species is similar to *P. depauperatum* but differs in being less densely tufted, in having more spikelets per panicle, and in bearing rounded spikelets. It is extremely close to *P. linearifolium,* but the panicles in *P. perlongum* are more than twice as long as broad, while in *P. linearifolium,* the panicles are broader. The grain also is considerably broader in *P. perlongum.*

15. Panicum linearifolium Scribn. in Britt. & Brown, Ill. Fl. 3:500. 1898.

Densely tufted perennial; culms erect or spreading, to 50 cm tall, glabrous to pilose; sheaths papillose-pilose or glabrous; ligule less than 1 mm long; blades 2–6 mm broad, scabrous or glabrous above, scabrous and puberulent or glabrous below; panicle to 10 cm long, ½–⅔ as broad, few-flowered, the flexuous branches ascending; spikelets 2.1–2.7 mm long, 1.2–1.5 mm broad, oblong-ellipsoid, obtuse to subacute, sparsely pubescent to glabrous;

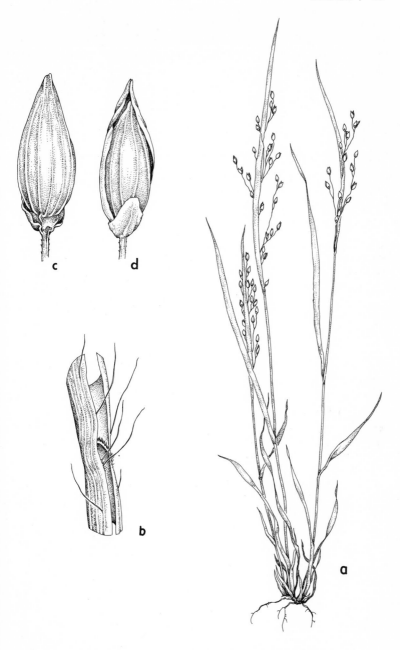

136. Panicum depauperatum (Panic Grass). *a.* Habit, X¾. *b.* Sheath, with ligule, X5. *c.* Spikelet, front view, X25. *d.* Spikelet, back view, X25.

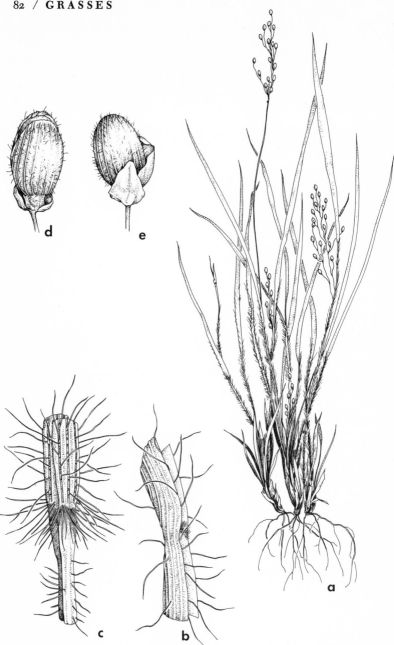

137. *Panicum perlongum* (Panic Grass). *a.* Habit, X½. *b.* Sheath, with ligule, X5. *c.* Sheath and node, X5. *d.* Spikelet, front view, X12½. *e.* Spikelet, back view, X12½.

second glume and sterile lemma equal, as long as the grain; grain 2.0–2.1 mm long, 1.2 mm broad, ovoid, obscurely umbonate; autumnal form similar, the panicle reduced.

Two varieties may be found in Illinois, distinguished as follows:

1. Sheaths papillose-pilose_____15a. *P. linearifolium* var. *linearifolium*
1. Sheaths glabrous_____15b. *P. linearifolium* var. *werneri*

15a. Panicum linearifolium Scribn. var. linearifolium *Fig. 138.*

Culms to 45 cm tall, glabrous; sheaths papillose-pilose; blades 2–4 mm broad, scabrous above, scabrous and puberulent below; spikelets 2.2–2.7 mm long, 1.3–1.5 mm broad, obtuse, sparsely pilose; first glume sparsely pubescent.

COMMON NAME: Panic Grass.

HABITAT: Dry woodlands.

RANGE: Quebec to Wisconsin and Iowa, south to Texas and Georgia.

ILLINOIS DISTRIBUTION: Occasional in the southern one-fourth of the state, rare throughout the remainder of the state.

Some specimens tend to resemble *P. depauperatum*, but this latter species has longer and acutely beaked spikelets. Both taxa often grow together in southern Illinois in dry woodlands.

15b. Panicum linearifolium Scribn. var. werneri (Scribn.)

Fern. Rhodora 23:194. 1921. *Fig. 139.*

Panicum werneri Scribn. in Britt. & Brown, Ill. Fl. 3:501. 1898.

Culms to 50 cm tall, sparsely pilose on the nodes; sheaths glabrous; blades to 6 mm broad, glabrous except for the scabrous, ciliate base; spikelets 2.1–2.4 mm long, 1.2–1.3 mm broad, obtuse to subacute, glabrous or puberulent; first glume glabrous or nearly so.

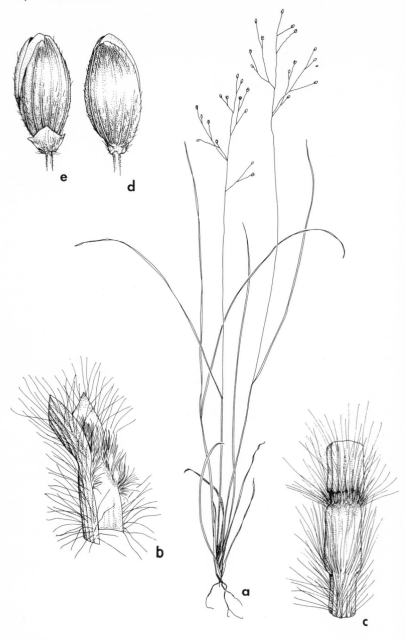

138. *Panicum linearifolium* var. *linearifolium* (Panic Grass). *a*. Habit,
X½. *b*. Sheath, with ligule, X7½. *c*. Sheath and node, X7½. *d*. Spikelet,
front view, X12½. *e*. Spikelet, back view, X12½.

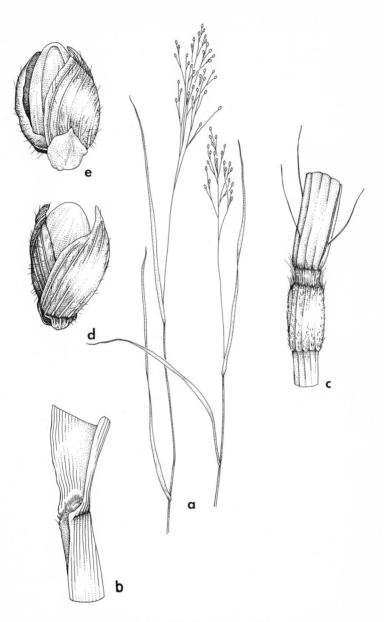

139. Panicum linearifolium var. *werneri* (Panic Grass). *a.* Upper part of plants, X½. *b.* Sheath, with ligule, X7½. *c.* Sheath and node, X7½. *d.* Spikelet, front view, X12½. *e.* Spikelet, back view, X12½.

COMMON NAME: Panic Grass.

HABITAT: Wooded slopes.

RANGE: Quebec to Minnesota, south to Texas and Virginia.

ILLINOIS DISTRIBUTION: Very rare; known only from LaSalle County, where it was collected in 1943.

This taxon, originally described as a distinct species and maintained in that rank by Hitchcock (1950), is considered here as a variety of *P. linearifolium* since the only difference is in the absence of pubescence on the sheath in var. *werneri*.

SECTION **Laxiflora**

16. Panicum laxiflorum Lam. Encycl. 4:748. 1798. *Fig. 140.*

Panicum xalapense HBK. Nov. Gen. & Sp. 1:103. 1815.

Tufted perennial; culms slender, erect, geniculate at the base, to 60 cm tall, glabrous except for the retrorsely pubescent nodes; sheaths glabrous or retrorsely pilose; ligule less than 1 mm long; blades to 10 mm broad, nearly glabrous to pilose on both surfaces, short-ciliate; panicle to 15 cm long, nearly as wide, rather few-flowered, the branches spreading or reflexed; spikelets 1.8–2.5 mm long, 1.0–1.1 mm broad, oblong-obovoid, obtuse, papillose-pubescent; first glume ¼–⅓ as long as the spikelet, subacute; second glume and sterile lemma equal, shorter than the grain; grain 1.5–1.7 mm long, 1.0–1.1 mm broad, broadly ellipsoid, minutely umbonate; autumnal form branching from the base, the blades much reduced; 2n = 18 (Brown, 1948).

COMMON NAME: Panic Grass.

HABITAT: Woodlands.

RANGE: Maryland to Missouri, south to Texas and Florida; Mexico; Central America.

ILLINOIS DISTRIBUTION: Common in the southern tip of the state; also in two central counties; apparently absent elsewhere.

Most recent Illinois botanists have called our specimens *P. xalapense*, while Fernald (1950) and Gleason (1952), among others, equate *P. xalapense* with *P. laxiflorum*. After studying specimens throughout the range of both taxa, I have concluded that there is no clear-cut demarcation separating the two.

The distinguishing features of *P. laxiflorum* are its pale green color, its long retrorse pilosity of the sheaths, and its long-ciliate blades. It is not particularly closely related to any other species of *Panicum* in Illinois.

Panicum laxiflorum often forms sterile tufts of densely long-ciliate blades during the autumn in dry oak-hickory upland woods of southern Illinois.

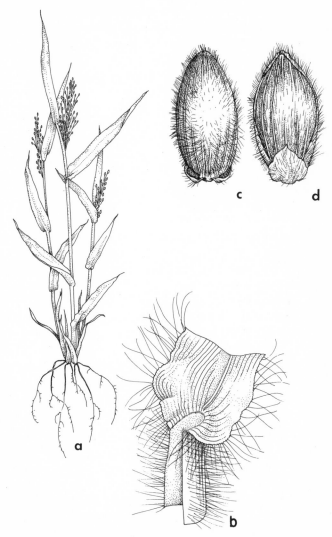

140. *Panicum laxiflorum* (Panic Grass). *a.* Habit, X½. *b.* Sheath, with ligule, X5. *c.* Spikelet, front view, X17½. *d.* Spikelet, back view, X17½.

SECTION **Dichotoma**

17. **Panicum microcarpon** Muhl. ex Ell. Bot. S.C. & Ga. 1:127. 1816. *Fig. 141.*

Panicum nitidum var. *ramulosum* Torr. Fl. North. & Mid. U. S. 146. 1824.

Cespitose perennial; culms erect or geniculate at the base, to 100 cm tall, the nodes with reflexed hairs; sheaths glabrous, ciliate, or the lowermost pubescent; ligule less than 1 mm long; blades thin, spreading, 8–15 mm broad, glabrous except for papillose cilia at the base; panicle to 12 cm long, many-flowered, the branches ascending; spikelets 1.5–1.6 mm long, 0.7 mm broad, ellipsoid, obtuse to subacute, glabrous or puberulent; first glume up to ¼ as long as the spikelet; second glume slightly shorter than the sterile lemma, barely exposing the grain at maturity; grain 1.3–1.4 mm long, 0.7 mm broad, ellipsoid, acute or subacute; autumnal form reclining, branching from all the nodes, blades reduced, the cilia very conspicuous, panicles much reduced, few-flowered; 2n = 18 (Brown, 1948).

COMMON NAME: Panic Grass.

HABITAT: Wet ground, often in woods.

RANGE: New Hampshire to Michigan and Missouri, south to Texas and Florida.

ILLINOIS DISTRIBUTION: Occasional in the southern one-third of the state; absent elsewhere, except for Peoria County. Gleason (1952) considers this species to be a variety of *P. nitidum* with shorter, glabrous spikelets, calling it var. *ramulosum*. *Panicum microcarpon* belongs to Section Dichotoma, along with *P. nitidum, P. boreale, P. dichotomum, P. mattamuskeetense,* and *P. yadkinense,* all species with longer and broader spikelets. The spikelets in *P. microcarpon* never exceed 1.6 mm in length and 0.7 mm in width.

Although most specimens from Illinois have glabrous spikelets, a few have puberulent ones. The blades vary in width from 8 to 15 millimeters.

The autumnal phase, because of its great branching, forms dense mats in low woods. On the other hand, the autumnal phase of *P. nitidum,* the most closely related species, is erect or ascending.

Panicum microcarpon begins to flower in mid-May.

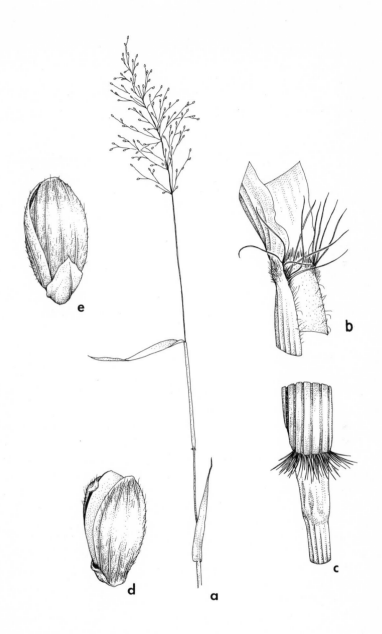

141. *Panicum microcarpon* (Panic Grass). *a.* Upper part of plant, X½. *b.* Sheath, with ligule, X7½. *c.* Sheath and node, X7½. *d.* Spikelet, front view, X20. *e.* Spikelet, back view, X20.

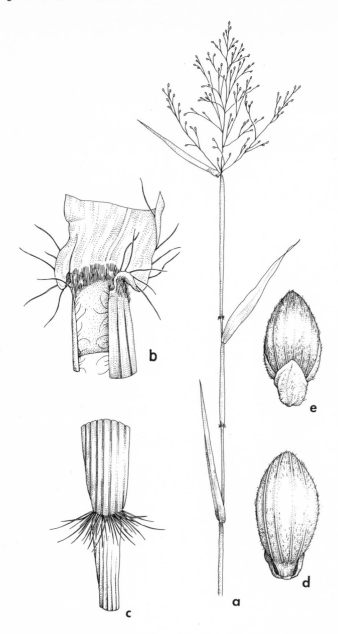

142. Panicum nitidum (Panic Grass). *b.* Upper part of plant, X½. *b.* Sheath, with ligule, X7½. *c.* Sheath and node, X7½. *d.* Spikelet, front view, X17½. *e.* Spikelet, back view, X17½.

18. Panicum nitidum Lam. Tabl. Encycl. 1:172. 1791. *Fig. 142.*

Panicum dichotomum var. *nitidum* (Lam.) Wood, Class-book 786. 1861.

Cespitose perennial; culms erect or spreading, to nearly 1 m tall, the nodes with reflexed hairs (except sometimes the uppermost); lower sheaths pubescent, the upper glabrous except for the ciliate margins, viscid-spotted; ligule less than 1 mm long; blades to 10 mm broad, glabrous except for the cilia near the base; panicle to 8 cm long, densely flowered, viscid, the branches ascending; spikelets 1.8–2.0 mm long, 1.0 mm broad, ellipsoid, obtuse, puberulent; first glume less than ⅓ as long as the spikelet, acute; second glume and sterile lemma subequal, barely as long as the grain; grain 1.7–1.8 mm long, 1.0 mm broad, broadly ellipsoid, obtuse to subacute; autumnal form erect to ascending, with reduced, involute blades and reduced, few-flowered panicles.

COMMON NAME: Panic Grass.

HABITAT: Xeric limestone bluff-top (in Illinois).

RANGE: Virginia to Florida and Texas; Illinois; Missouri; West Indies; Mexico.

ILLINOIS DISTRIBUTION: Very rare; known only from Jackson County (Devil's Backbone, one mile north of Grand Tower, July 3, 1963, *J. Ozment 960*).

This species differs from *P. microcarpon* in its longer, hairier spikelets and from *P. dichotomum* in its more abundant pubescence.

Panicum nitidum is primarily a southern species with single stations in Missouri and Illinois. Where it is common in the southeast, it usually occupies moist, sandy areas.

19. Panicum boreale Nash, Bull. Torrey Club 22:421. 1895. *Fig. 143.*

Cespitose perennial; culms erect, to 50 cm tall, the nodes glabrous or sparsely pilose; sheaths ciliate on the margins, or the lower sometimes puberulent throughout; ligules less than 0.5 mm long; blades to 12 mm broad, sparsely ciliate at the base, otherwise glabrous; panicle to 10 cm long, ¾ as broad, few-flowered, the branches ascending to spreading; spikelets 2.0–2.2 mm long, 1.0 mm broad, ellipsoid, subacute, pubescent; first glume up to ⅓ as long as the spikelet, acute, minutely pubescent to nearly gla-

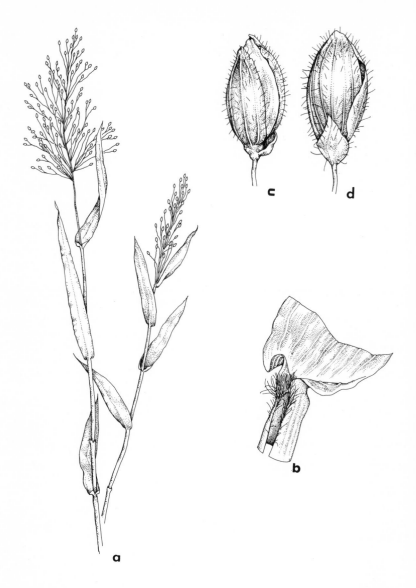

143. Panicum boreale (Northern Panic Grass). *a.* Upper part of plants, X½. *b.* Sheath, with ligule, X5. *c.* Spikelet, front view, X15. *d.* Spikelet, back view, X15.

brous; second glume and sterile lemma subequal, nearly as long as the grain; grain 1.9 mm long, 1.0 mm broad, ellipsoid, subacute; autumnal form erect, sparsely branched, the leaves and panicles scarcely reduced.

COMMON NAME: Northern Panic Grass.

HABITAT: Moist sand.

RANGE: Newfoundland to Minnesota, south to Illinois and New Jersey.

ILLINOIS DISTRIBUTION: Very rare; known only from Lake County, where it was collected in 1944 by Fuller and Graham. This species somewhat resembles the rare pubescent-spikelet form of *P. dichotomum,* but *P. boreale* has the second glume and sterile lemma as long as the grain.

This is an enigmatic species with few clear-cut characters. More discriminating collecting in the extreme northern counties of the state may turn up more records for this species.

20. Panicum dichotomum L. Sp. Pl. 58. 1753.

Tufted perennial from a knotted crown; culms often purplish, erect, to 80 cm tall, the nodes glabrous or sparsely pilose, or the lowermost bearded; sheaths glabrous except for the sometimes ciliate margins, or the lowermost puberulent to softly pubescent; ligule less than 1 mm long; blades to 10 mm broad, ciliate on the margins and near the base, otherwise glabrous; panicle to 12 cm long, the branches spreading or drooping; spikelets 1.9–2.2 mm long, 0.9–1.0 mm broad, ellipsoid to ovoid, obtuse to subacute, glabrous or (rarely) pubescent; first glume ¼–⅓ as long as the spikelet, subacute to acute; second glume and sterile lemma subequal to equal, shorter than to as long as the grain; grain 1.8 mm long, 0.9–1.0 mm broad, ellipsoid, subacute to obscurely apiculate; autumnal form erect or reclining, branched from the middle nodes, with few or no leaves on the lower half of the culm, blades much reduced, numerous, often involute.

Two varieties, sometimes treated as species, may be recognized in Illinois.

1. Nodes glabrous or sparsely pubescent, not bearded; grain slightly exserted_____20a. *P. dichotomum* var. *dichotomum*
1. At least the lowermost nodes bearded; grain included_____
_____20b. *P. dichotomum* var. *barbulatum*

20a. Panicum dichotomum L. var. dichotomum *Fig. 144.*

Culms to 50 cm tall, the nodes glabrous or sparsely pilose; sheaths glabrous except for the sometimes ciliate margins, or the lowermost puberulent; blades to 8 mm broad; panicle to 10 cm long, the branches spreading; spikelets 1.9–2.0 mm long, 0.9 mm broad, ellipsoid, glabrous; first glume ⅓ as long as the spikelet, subacute; second glume and sterile lemma subequal, shorter than the grain; grain 0.9 mm broad, subacute; autumnal form erect; 2n = 18 (Brown, 1948).

COMMON NAME: Panic Grass.
HABITAT: Dry soil, usually in woodlands.
RANGE: New Brunswick to Illinois, south to Texas and Florida.
ILLINOIS DISTRIBUTION: Rather common in the southern one-fourth of the state, occasional to rare elsewhere. Reports of this taxon from Cook and Winnebago counties could not be verified.

Most Illinois specimens of *P. dichtotomum* var. *dichotomum* have glabrous spikelets, although a few specimens possess minutely pubescent spikelets.

The autumnal form, with few or no leaves on the lower half of the culm, was described by Linnaeus as resembling diminutive trees.

20b. Panicum dichotomum L. var. barbulatum (Michx.)
Wood, Class-book 786. 1861. *Fig. 145.*

Panicum barbulatum Michx. Fl. Bor. Am. 1:49. 1803.

Panicum pubescens var. *barbulatum* (Michx.) Britt. Cat. Pl. N. J. 280. 1889.

Culms slender, to 80 cm tall, the lower nodes bearded; upper sheaths glabrous, except for a puberulent ring at the summit, the lower sheaths softly pubescent; blades to 10 mm broad; panicle to 12 cm long, the branches spreading or drooping; spikelets 1.9–2.2 mm long, 0.9–1.0 mm broad, ovoid, glabrous; first glume ¼–⅓ as long as the spikelet, acute; second glume and sterile lemma equal, as long as the grain; grain 1.0 mm broad, obscurely apiculate; autumnal form much branched, reclining, with horizontally spreading blades; 2n = 18 (Brown, 1948).

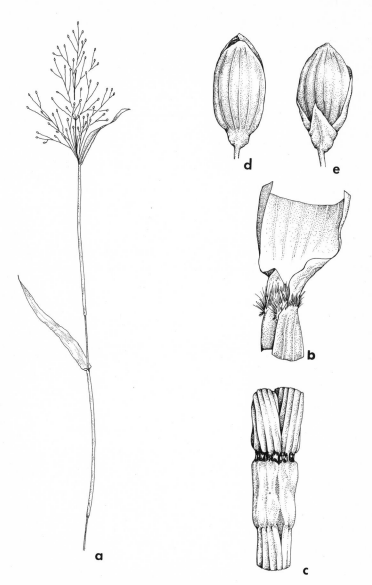

144. Panicum dichotomum var. *dichotomum* (Panic Grass).   *a.* Upper part of plant, X½.   *b.* Sheath, with ligule, X7½.   *c.* Sheath and node, X7½.   *d.* Spikelet, front view, X15.   *e.* Spikelet, back view, X15.

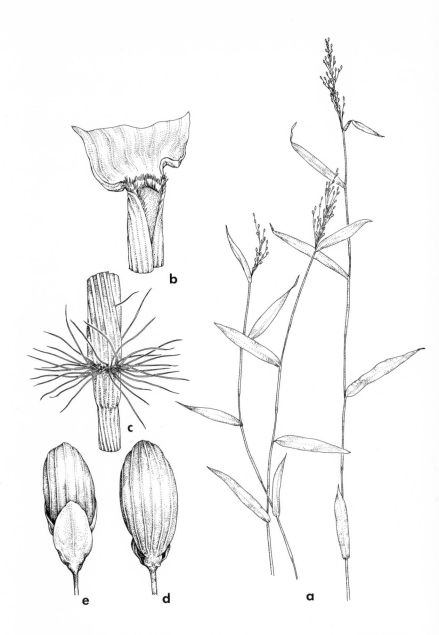

145. Panicum dichotomum var. *barbulatum* (Panic Grass). *a.* Upper part of plants, X½. *b.* Sheath, with ligule, X7½. *c.* Sheath and node, X7½. *d.* Spikelet, front view, X15. *e.* Spikelet, back view, X15.

COMMON NAME: Panic Grass.

HABITAT: Dry, usually rocky, woods.

RANGE: New Hampshire to Michigan and Illinois, south to Texas and Georgia.

ILLINOIS DISTRIBUTION: Occasional, but apparently not often collected.

Both varieties of *P. dichotomum* have extremely similar vernal forms, but are remarkably different in their autumnal aspects. In weighing all characters, I must conclude that varietal rank is the better for these two taxa.

21. Panicum mattamuskeetense Ashe, Jour. Elisha Mitchell Sci. Soc. 15:45. 1898. *Fig. 146.*

Sparsely tufted perennial; culms erect, to 1 m tall, glabrous except for the lowermost bearded nodes; lower sheaths velvety, the upper more or less glabrous; ligule less than 1 mm long; blades to 12 mm broad, usually velvety-pubescent or the uppermost sometimes glabrous; panicle to 12 cm long, the branches ascending to spreading; spikelets 2.2–2.7 mm long, 1.0–1.4 mm broad, ellipsoid, obtuse, pubescent; first glume about ¼ as long as the spikelet, pubescent; second glume and sterile lemma subequal; grain 2.0–2.5 mm long, ellipsoid; autumnal form sparingly branched, mostly erect.

COMMON NAME: Panic Grass.

HABITAT: Along a levee (in Illinois).

RANGE: Massachusetts to South Carolina; Indiana; Illinois.

ILLINOIS DISTRIBUTION: Known only from a single station in extreme southern Illinois (Massac County: Mermet Wildlife Refuge, May 17, 1966, *J. Schwegman 365*).

At the Illinois station, this grass was growing with *Wisteria macrostachya.*

The range for this species is the Coastal Plain from Massachusetts to South Carolina, as well as a single station in southern Indiana.

Distinguishing characteristics of *P. mattamuskeetense* are the bearded nodes, the usually velvety-pubescent leaves, and the pubescent spikelets averaging 2.5 mm long.

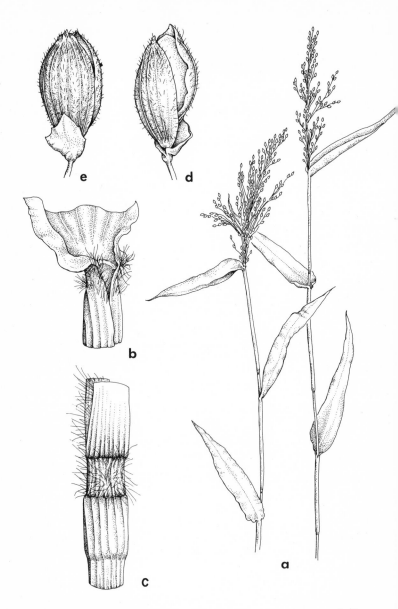

146. *Panicum mattamuskeetense* (Panic Grass). *a.* Upper part of plants, X½. *b.* Sheath, with ligule, X5. *c.* Sheath and node, X5. *d.* Spikelet, front view, X12½. *e.* Spikelet, back view, X12½.

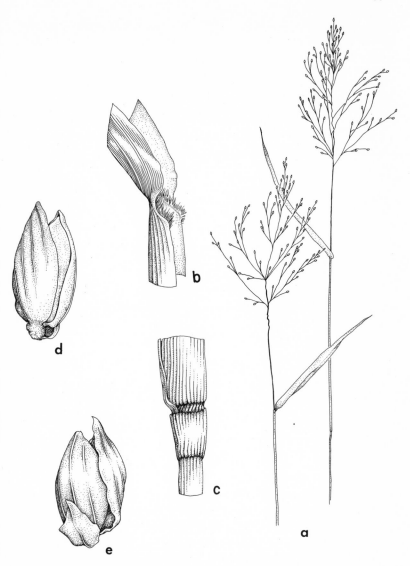

147. Panicum yadkinense (Panic Grass). *a.* Upper part of plants, X½. *b.* Sheath, with ligule, X7½. *c.* Sheath and node, X7½. *d.* Spikelet, front view, X12½. *e.* Spikelet, back view, X12½.

22. Panicum yadkinense Ashe, Jour. Elisha Mitchell Sci. Soc. 16:85. 1900. *Fig. 147.*

Tufted perennial from a knotted crown; culms erect, to nearly 1 m tall, the nodes glabrous or sparsely pilose; sheaths glabrous, with pale, glandular spots; ligule less than 1 mm long; blades to 12 mm broad, glabrous; panicle to 12 cm long, about ¾ as wide, the branches ascending; spikelets 2.3–2.5 mm long, 1.0 mm broad, ellipsoid, acute, glabrous; first glume ⅓ as long as the spikelet, obtuse to subacute, glabrous; second glume and sterile lemma equal, longer than the grain; grain 1.9 mm long, 0.9 mm broad, ellipsoid, obtuse to subacute; autumnal form more or less erect, loosely branching, the blades only slightly smaller than the vernal ones; 2n = 18 (Brown, 1948).

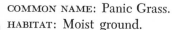

COMMON NAME: Panic Grass.

HABITAT: Moist ground.

RANGE: Pennsylvania to Michigan and southern Illinois, south to Texas and Georgia.

ILLINOIS DISTRIBUTION: Rare; restricted to a few southern counties.

This species is related most closely to *P. dichotomum,* but differs in its longer spikelets and its taller culms. The glandular spots on the sheaths are very distinctive.

Schneck's collection of this species from Tunnel Hill, Johnson County, in 1902, was apparently the first for this grass in Illinois.

SECTION **Lanuginosa**

23. **Panicum meridionale** Ashe, Journ. Elisha Mitchell Sci. Soc. 15:59. 1898.

Tufted perennial; culms slender, erect or ascending, appressed-pubescent, or the upper portion nearly glabrous; upper sheaths appressed-puberulent to grayish-villous, the lower papillose-pilose; ligule 2–5 mm long; blades 2–6 mm broad, spreading or ascending, long-pilose to puberulent above, glabrous to puberulent to grayish-villous beneath; panicle to 5 cm long, nearly as wide, loosely flowered, the axis glabrous or puberulent, the branches ascending; spikelets 1.3–1.5 mm long, 0.8–0.9 mm broad, obovoid to broadly ellipsoid, obtuse, minutely papillose-pubescent or pilose; first glume ¼–⅜ as long as the spikelet, acute to subacute, papillose-pubescent; second glume and sterile lemma equal or subequal, usually as long as the grain; grain 1.2–1.3 mm long, 0.8–0.9 mm broad, broadly ellipsoid, obscurely pointed or obtuse; autumnal form nearly erect to spreading, branching from all the nodes, sometimes forming mats, the blades and panicles more or less reduced.

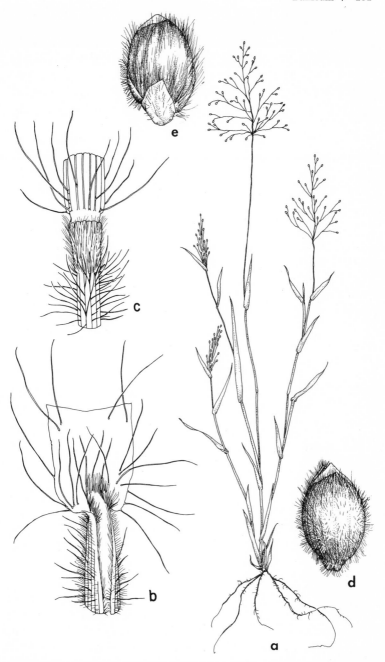

148. *Panicum meridionale* var. *meridionale* (Panic Grass). *a*. Habit, X½.
b. Sheath, with ligule, X5. *c*. Sheath and node, X5. *d*. Spikelet, front view,
X20. *e*. Spikelet, back view, X20.

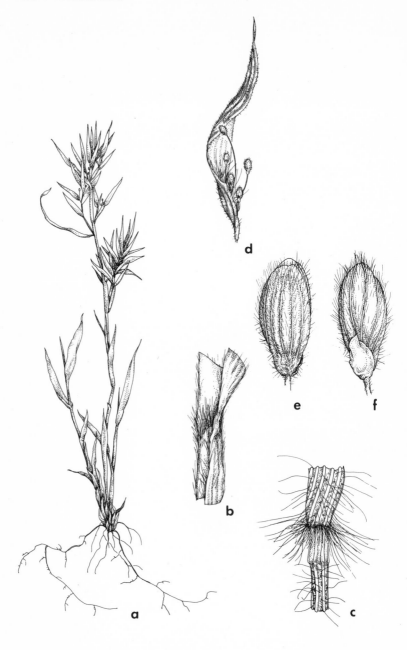

149. *Panicum meridionale* var. *albemarlense* (Panic Grass). *a*. Habit, X½.
b. Sheath, with ligule, X5. *c*. Sheath and node, X5. *d*. Inflorescence, X2½.
e. Spikelet, front view, X20. *f*. Spikelet, back view, X20.

Two varieties of *P. meridionale* occur in Illinois. There may be some justification in maintaining them as distinct species, as has been done by Hitchcock and others.

1. Vernal blades 1.5–4.0 cm long; plants greenish-yellow; autumnal form nearly erect_____23a. *P. meridionale* var. *meridionale*
1. Vernal blades 4.5–7.0 cm long; plants grayish; autumnal form spreading to ascending, but eventually forming mats_____
_____23b. *P. meridionale* var. *albemarlense*

23a. Panicum meridionale Ashe var. meridionale *Fig. 148.*

Panicum unciphyllum meridionale (Ashe) Scribn. & Merrill, Rhodora 3:123. 1901.

Culms appressed-pubescent, or the upper part nearly glabrous; upper sheaths appressed-pubescent, the lower papillose-pilose; blades 2–4 mm broad, long-pilose to nearly glabrous above, glabrous to puberulent beneath; spikelets obovoid, minutely papillose-pubescent; first glume ¼–⅓ as long as the spikelet; autumnal form nearly erect.

COMMON NAME: Panic Grass.

HABITAT: Sandy woodlands.

RANGE: Nova Scotia to Minnesota, south to Alabama and Georgia.

ILLINOIS DISTRIBUTION: Rare; known from the extreme northeastern counties and two southern counties.

This variety differs from var. *albemarlense* in its vernal phase by its yellow-green color, its shorter leaves, and its appressed-pubescent stems. In the autumnal forms, var. *meridionale* tends to be nearly erect, while var. *albemarlense* ultimately is mat-forming.

Panicum meridionale belongs to Section Lanuginosa by virtue of its ligules which are composed of a ring of hairs up to 5 mm long.

23b. Panicum meridionale Ashe var. albemarlense (Ashe) Fern. Rhodora 36:76. 1934. *Fig. 149.*

Panicum albemarlense Ashe, Journ. Elisha Mitchell Sci. Soc. 16:84. 1900.

Culms grayish-villous; sheaths grayish-villous, often papillose-pilose; blades to 6 mm broad, the upper surface puberulent, the lower surface grayish-villous; spikelets broadly ellipsoid, pilose; first glume ⅔ as long as the spikelet; autumnal form spreading to ascending, much branched, eventually forming mats, the blades reduced and exceeding the panicles.

COMMON NAME: Panic Grass.

HABITAT: Margins of wet, peaty meadow (in Illinois).

RANGE: Massachusetts to Minnesota, south to Tennessee and North Carolina; Illinois.

ILLINOIS DISTRIBUTION: Very rare; known only from Kankakee County (Saline Township, September 6, 1936, *R. A. Schneider 147*).

I am following Fernald (1950) in considering *P. albemarlense* as a variety of *P. meridionale,* although Gleason (1952) combines it with that species. It differs in its grayish pubescence and its mat-forming autumnal form.

24. Panicum lanuginosum Ell. Bot. S. C. & Ga. 1:123. 1816.

Tufted perennial; culms spreading to ascending to stiffly erect, to 80 cm tall, spreading papillose-pilose throughout, or glabrous above and pilose below, or (rarely) glabrous throughout; sheaths papillose-pilose or (rarely) merely with ciliate margins; ligule 3–5 mm long; blades to 12 mm broad, variously pubescent, the hairs up to 6 mm long and often papillose; panicle to 10 cm long, nearly as wide, loosely to densely flowered, the axis glabrous, appressed-pubescent, or long-pilose, the lower branches spikelets 1.3–1.9 mm long, 0.8–1.0 mm broad, ellipsoid to obovoid, obtuse to subacute, papillose-pubescent; first glume ¼–⅓ as long as the spikelet, truncate to obtuse to subacute; second glume and sterile lemma equal to subequal, shorter than to as long as the grain; grain 1.2–1.6 mm long, 0.8–1.0 mm broad, ellipsoid, obtuse to subacute to obscurely apiculate; autumnal form erect or spreading or decumbent and mat-forming, loosely branched, the blades reduced, often involute at the tip.

This is one of the most difficult complexes of plants in Illinois. Four varieties are recognized as occurring in Illinois, although each in its extreme condition is quite different from the others and has been recognized as a distinct species by several authors. The typical var. *lanuginosum* is a more southern taxon which has not as yet been found in Illinois.

The four taxa in Illinois may be separated as follows:

1. Axis of panicle branches with some long-spreading hairs.

2. Upper surface of blades glabrous except for long hairs at the base_____24a. *P. lanuginosum* var. *fasciculatum*
2. Upper surface of blades pilose or appressed-pubescent.
　3. Spikelets 1.6–1.9 mm long; lower panicle branches ascending _____24a. *P. lanuginosum* var. *fasciculatum*
　3. Spikelets 1.3–1.5 mm long; lower panicle branches drooping and entangled_____24b. *P. lanuginosum* var. *implicatum*
1. Axis of panicle branches either glabrous or with appressed hairs.
　4. Spikelets 1.4–1.6 mm long_____ _____24c. *P. lanuginosum* var. *lindheimeri*
　4. Spikelets 1.6–1.9 mm long_____ _____24d. *P. lanuginosum* var. *septentrionale*

24a. Panicum lanuginosum Ell. var. **fasciculatum** (Torr.) Fern. Rhodora 36:77. 1934. *Fig. 150.*

Panicum dichotomum var. *fasciculatum* Torr. Fl. North. & Mid. U. S. 145. 1824.
Panicum nitidum var. *ciliatum* Torr. Fl. North. & Mid. U. S. 146. 1824.
Panicum nitidum var. *pilosum* Torr. Fl. North. & Mid. U. S. 146. 1824.
Panicum huachucae Ashe, Journ. Elisha Mitchell Sci. Soc. 15:51. 1898.
Panicum tennesseense Ashe, Journ. Elisha Mitchell Sci. Soc. 15:52. 1898.
Panicum lanuginosum var. *huachucae* (Ashe) Hitchc. Rhodora 8:208. 1906.
Panicum huachucae var. *silvicola* Hitchc. & Chase in Robins. Rhodora 10:64. 1908.
Panicum huachucae var. *fasciculatum* (Torr.) Fern. Rhodora 14:171. 1912.
Panicum lindheimeri var. *fasciculatum* (Torr.) Fern. Rhodora 23:228. 1921.
Panicum lindheimeri var. *tennesseense* (Ashe) Farw. Am. Midl. Nat. 11:45. 1928.
Panicum lanuginosum var. *tennesseense* (Ashe) Gl. Phytologia 4:21. 1952.

Tufted perennial; culms spreading to ascending to stiffly erect, green to purplish, to 75 cm tall, papillose-pilose to glabrous, the nodes retrorsely bearded; sheaths papillose-pilose, rarely glabrous; blades to 12 mm broad, densely pubescent to nearly glabrous above, short papillose-pilose or nearly glabrous beneath; panicle to nearly 10 cm long, nearly as wide, the axis pilose, the branches spreading to ascending; spikelets 1.6–1.9 mm long, 0.8–1.0 mm broad, ellipsoid to obovoid, obtuse, the pubescence often papillose; first glume ¼–⅓ as long as the spikelet, obtuse to subacute; second glume and sterile lemma equal or subequal, shorter than to as long as the grain; grain 1.4–1.6 mm long, 0.8–1.0 mm broad, obtuse to subacute to obscurely apiculate; autumnal form widely spreading or decumbent, often forming prostrate mats, the blades much reduced and ciliate near the base; 2n = 18 (Brown, 1948, for *P. huachucae* and *P. tennesseense*).

COMMON NAME: Panic Grass.

HABITAT: Usually low, moist, open situations; also in thin, dry woods.

RANGE: Nova Scotia to North Dakota and Utah, south to Arizona, Texas, and Florida.

ILLINOIS DISTRIBUTION: Common throughout the state.

As treated here this taxon includes those plants in Illinois referred previously to *P. huachucae,* as well as those specimens of *P. tennesseense* which have pilose panicle branches. Differences between these latter two are virtually lacking. The character usually used is the glabrous upper leaf surface of *P. tennesseense* as opposed to the short papillose-pilose upper leaf surface of *P. huachucae.* Study of only a few specimens of this complex, however, shows that this character hopelessly overlaps and is unreliable.

The relationship between var. *fasciculatum* and var. *implicatum* is close, but the latter has shorter spikelets and drooping, entangled lower panicle branches.

24b. Panicum lanuginosum Ell. var. implicatum (Scribn.)
Fern. Rhodora 36:77. 1934. *Fig. 151.*

Panicum implicatum Scribn. in Britt. & Brown, Ill. Fl. 3:498. 1898.

Panicum unciphyllum implicatum (Scribn.) Scribn. & Merrill, Rhodora 3:123. 1901.

Culms slender, erect or ascending, to 60 cm tall, papillose-pilose; sheaths papillose-pilose; blades to 6 mm broad, the upper surface pilose with hairs 3–4 mm long, the lower surface papillose-pubescent, the hairs more or less appressed; panicle to 6 cm

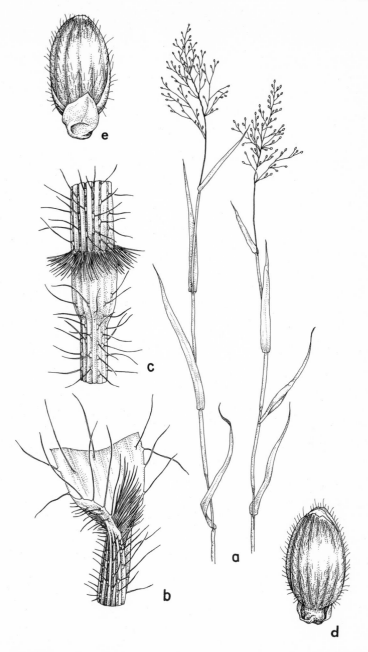

150. *Panicum lanuginosum* var. *fasciculatum* (Panic Grass). *a*. Upper part of plants, X½. *b*. Sheath, with ligule, X7½. *c*. Sheath and node, X7½. *d*. Spikelet, front view, X20. *e*. Spikelet, back view, X20.

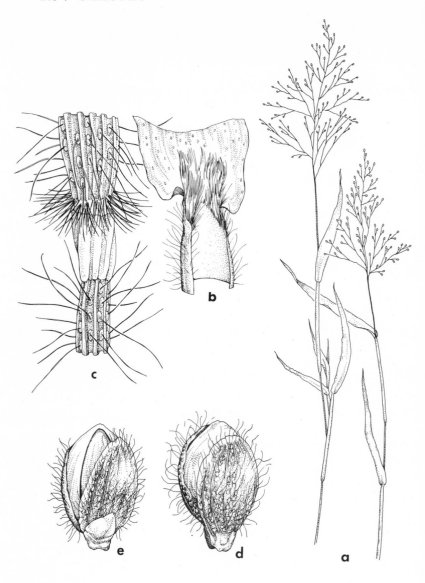

151. *Panicum lanuginosum* var. *implicatum* (Panic Grass). *a.* Upper part of plants, X½. *b.* Sheath, with ligule, X7½. *c.* Sheath and node, X7½. *d.* Spikelet, front view, X20. *e.* Spikelet, back view, X20.

long, nearly as wide, the axis long-pilose, the lower branches drooping and often tangled; spikelets 1.3–1.5 mm long, 0.9 mm broad, obovoid, obtuse, papillose-pilose; first glume ¼ as long as the spikelet, obtuse, papillose-pubescent; second glume and sterile lemma equal, as long as the grain; grain 1.2–1.3 mm long, 0.8–0.9 mm broad, obtuse, minutely umbonate; autumnal form spreading to erect, loosely branched, geniculate below, the blades and panicles reduced.

COMMON NAME: Panic Grass.

HABITAT: Swampy soil or in moist depressions of sandstone cliffs.

RANGE: Newfoundland to Minnesota, south to Missouri, Tennessee, and Virginia.

ILLINOIS DISTRIBUTION: Occasional throughout the state. In its extreme condition, var. *implicatum* is very distinctive in appearance with its drooping, entangled lower panicle branches and its narrow, long-pilose blades.

Too much intergradation, however, may be found between var. *implicatum* and var. *fasciculatum* to make them specifically distinct.

24c. **Panicum lanuginosum** Ell. var. **lindheimeri** (Nash) Fern. Rhodora 36:77. 1934. *Fig. 152.*

Panicum lindheimeri Nash, Bull. Torrey Club 24:196. 1897.
Panicum lindheimeri var. *typicum* Fern. Rhodora 23:227. 1921.

Culms stiffly ascending, to 80 cm tall, glabrous above, ascending-pubescent below; sheaths ciliate on the margins, the lower with ascending pubescence; blades to 9 mm broad, glabrous above, glabrous or puberulent below, papillose-ciliate at the base; panicles to 7.5 cm long, nearly as wide, the branches ascending to spreading; spikelets 1.4–1.6 mm long, 0.8–0.9 mm broad, obovoid, obtuse to subacute, pubescent; first glume up to ¼ as long as the spikelet, obtuse; second glume and sterile lemma equal, a little shorter than the grain; grain 1.3–1.4 mm long, 0.8 mm broad, obtuse to subacute; autumnal form stiffly spreading or prostrate, the blades reduced and often involute at the tip; 2n = 18 (Brown, 1948, as *P. lindheimeri*).

COMMON NAME: Panic Grass.

HABITAT: Sandy soil, usually in woodlands.

RANGE: Quebec to Minnesota, south to Kansas, New Mexico, Texas, and Florida; California.

ILLINOIS DISTRIBUTION: Occasional throughout the state. There is considerable variation in degree of pubescence or lack of it of the culms, sheaths, and blades.

This taxon, which differs from var. *fasciculatum* and var. *implicatum* primarily by its nearly glabrous panicle branches, has been considered by several authors to be a distinct species. In fact, Hitchcock has assigned this taxon to a different section, Section Spreta, on the basis of its general lack of culm and leaf pubescence. It is my opinion that Section Spreta, set up on such a tenuous character as vegetative pubescence, should not be distinguished from Section Lanuginosa.

The relatively small spikelets are nearly identical with those of var. *implicatum*.

24d. **Panicum lanuginosum** Ell. var. **septentrionale** (Fern.) Fern. Rhodora 36:77. 1934. *Fig. 153.*

Panicum lindheimeri var. *septentrionale* Fern. Rhodora 23:227. 1921.

Tufted perennial; culms erect to spreading, green to purplish, to 60 cm tall, usually glabrous but occasionally sparsely papillose-pilose; sheaths mostly glabrous; blades to 10 mm broad, glabrous above except for long hairs at the base, the lower surface usually nearly glabrous; panicle to 8 cm long, nearly as wide, the axis glabrous, the branches ascending; spikelets 1.6–1.9 mm long, 0.8–1.0 mm broad, obovoid, obtuse, pubescent; first glume ¼ as long as the spikelet, subacute to nearly truncate; second glume and sterile lemma nearly equal, shorter than the grain; grain 1.4–1.5 mm long, 0.8 mm broad, obtuse to subacute.

COMMON NAME: Panic Grass.

HABITAT: Moist soil.

RANGE: Nova Scotia to Manitoba, south to Virginia and Kansas.

ILLINOIS DISTRIBUTION: Collections from Jefferson and Johnson counties are the only ones I have seen during the preparation of this book, although I suspect this variety is more widespread in Illinois.

There may be some question about the recognition of

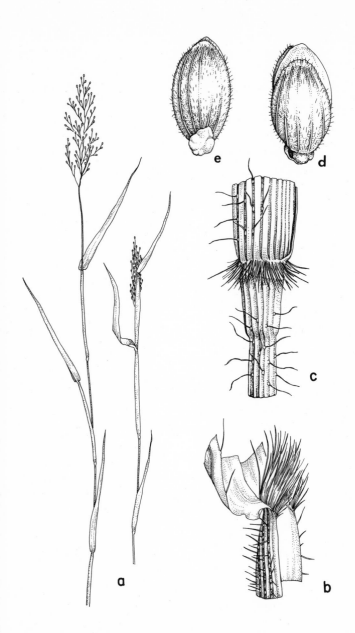

152. *Panicum lanuginosum* var. *lindheimeri* (Panic Grass). *a*. Upper part of plants, X½. *b*. Sheath, with ligule, X2½. *c*. Sheath and node, X7½. *d*. Spikelet, front view, X20. *e*. Spikelet, back view, X20.

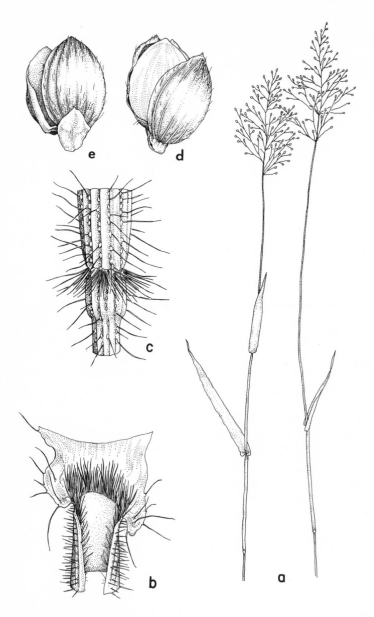

153. *Panicum lanuginosum* var. *septentrionale* (Panic Grass). *a*. Upper part of plants, X½. *b*. Sheath, with ligule, X7½. *c*. Sheath and node, X7½. *d*. Spikelet, front view, X20. *e*. Spikelet, back view, X20.

this variety from var. *lindheimeri* in that the only major difference lies in the size of the spikelets.

25. Panicum praecocius Hitchc. & Chase, Rhodora 8:206. 1906. *Fig. 154.*

Tufted perennial; culms very slender, wiry, erect to spreading, geniculate at the base, to 25 cm tall, with papillose pubescence horizontally spreading, to 4 mm long; sheaths papillose-pilose; ligule 3–4 mm long; blades to 7 mm broad, erect to ascending, rather firm, the upper surface pilose with hairs to 5 mm long, the lower surface papillose-pilose; panicle to 8 cm long, nearly as wide, the axis pilose, the branches spreading to ascending; spikelets 1.8–1.9 mm long, 1.0 mm broad, obovoid, obtuse, pilose; first glume ⅓–½ as long as the spikelet, triangular, acute; second glume and sterile lemma subequal, shorter than the grain; grain 1.6 mm long, 1.0 mm broad, broadly ellipsoid, subacute; autumnal form ascending to erect, the blades scarcely reduced and much exceeding the reduced panicles.

COMMON NAME: Panic Grass.
HABITAT: Dry soil, often in prairies.
RANGE: Michigan to North Dakota, south to Texas and Arkansas.
ILLINOIS DISTRIBUTION: Occasional throughout the state; rare in the extreme southern counties.

The type was collected in Stark County, Illinois, by V. H. Chase in 1900.

Several characters separate this species from the many variations of *P. lanuginosum*. These are the horizontally spreading long hairs of the culm and the very long hairs over the entire upper surface of the leaves, and the relative longer length of the first glume in comparison with the length of the spikelet.

26. Panicum subvillosum Ashe, Journ. Elisha Mitchell Sci. Soc. 16:86. 1900. *Fig. 155.*

Tufted perennial; culms slender, spreading to ascending, to 50 cm tall, ascending papillose-pilose, the nodes short-bearded; sheaths sparsely ascending papillose-pilose; ligule 2.5–3.5 mm long; blades to 6 mm broad, ascending, rather firm, pilose on both surfaces with hairs 3–5 mm long; panicle to 5 cm long, ⅔–¾ as wide, densely-flowered, the axis pilose, the branches as-

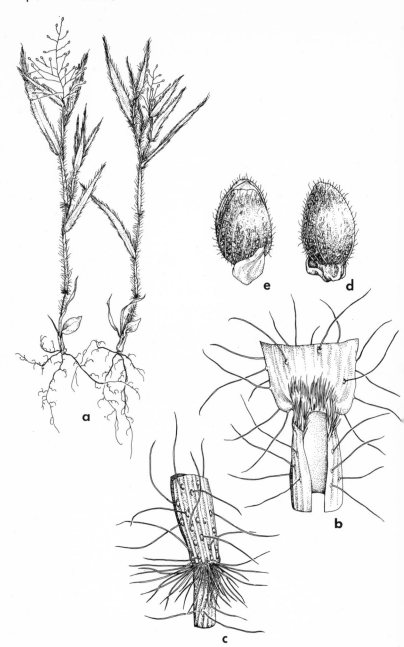

154. *Panicum praecocius* (Panic Grass). *a*. Habit, X½. *b*. Sheath, with ligule, X7½. *c*. Sheath and node, X7½. *d*. Spikelet, front view, X12½. *e*. Spikelet, back view, X12½.

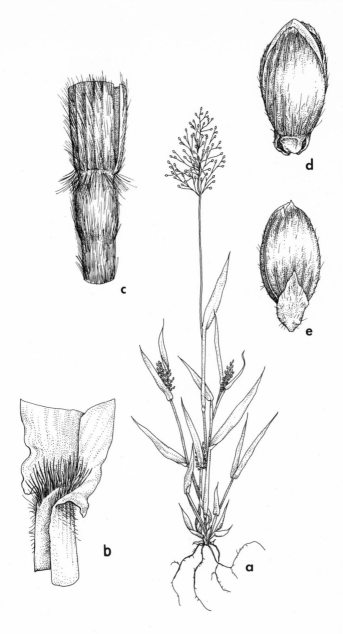

155. *Panicum subvillosum* (Panic Grass). *a.* Habit, X½. *b.* Sheath, with ligule, X5. *c.* Sheath and node, X5. *d.* Spikelet, front view, X17½. *e.* Spikelet, back view, X17½.

cending; spikelets 1.6–1.9 mm long, 0.9 mm broad, ellipsoid, obtuse, pubescent; first glume about ½ as long as the spikelet, acute to acuminate; second glume and sterile lemma equal, a little shorter than the grain; grain 1.4–1.5 mm long, 0.8 mm broad, ellipsoid, obtuse; autumnal form spreading, sparsely branched, the blades and panicles only moderately reduced; 2n = 18 (Church, 1929).

COMMON NAME: Panic Grass.

HABITAT: Sandy soil in an open field (in Illinois).

RANGE: Nova Scotia to Minnesota, south to Missouri, northwestern Indiana, and Rhode Island.

ILLINOIS DISTRIBUTION: Very rare; known only from Lake County (north of Barrington, August 4, 1964, *J. Ozment 2862*).

Kibbe (1952) reported this species from Hancock County, but no specimen could be found to substantiate this.

Panicum subvillosum has longer spikelets than either *P. meridionale* or *P. lanuginosum* var. *implicatum* and has longer pilosity on the blades than *P. lanuginosum* var. *fasciculatum*. It differs from *P. praecocius* by the sparse, ascending hairs on the culm, by its narrower panicles, and by the absence of reduced panicles in the autumnal phase.

27. **Panicum villosissimum** Nash, Bull. Torrey Club 23:149. 1896.

Densely tufted perennial; culms slender, erect or ascending, to 50 cm tall, olive-green, papillose-pilose; sheaths papillose-pilose; ligule 2–5 mm long; blades to 10 mm broad, rather firm, ascending to spreading, the upper surface glabrous to short-pilose to appressed long-pilose, the lower surface pilose; panicle to 12 cm long, nearly as wide, loosely flowered, the axis sparsely pilose, the branches spreading to ascending; spikelets 2.0–2.4 mm long, 1.1 mm broad, oblong-ellipsoid, obtuse to subacute, papillose-pubescent; first glume ⅓–½ as long as the spikelet, acute, glabrous or nearly so; second glume and sterile lemma subequal, slightly shorter than the grain; grain 1.9 mm long, ellipsoid, subacute; autumnal form spreading to prostrate, sparsely branching, the blades more or less reduced and often glabrous above.

Two varieties occur in Illinois.

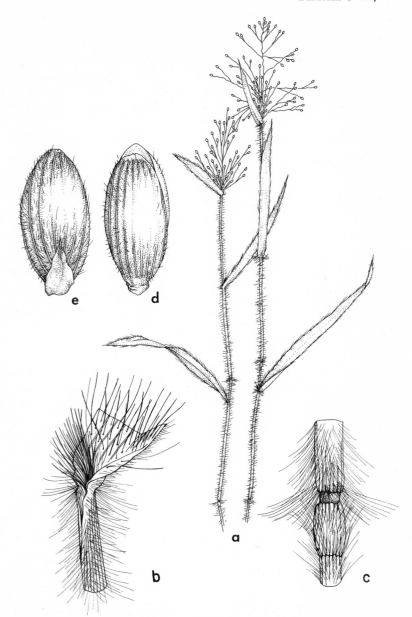

156. *Panicum villosissimum* var. *villosissimum* (Hairy Panic Grass). *a.* Upper part of plants, X½. *b.* Sheath, with ligule, X5. *c.* Sheath and node, X5. *d.* Spikelet, front view, X17½. *e.* Spikelet, back view, X17½.

1. Culms and sheaths spreading papillose-pilose; ligule 4–5 mm long; upper surface of blades appressed long-pilose_ _ _ _ _ _ _ _ _ _ _ _ _ _ _ _ _
_ _27a. *P. villosissimum* var. *villosissimum*
1. Culms and sheaths appressed papillose-pilose; ligule 2–3 mm long; upper surface of blades glabrous or short-pilose_ _ _ _ _ _ _ _ _ _ _ _ _ _
_ _ _ _ _ _ _ _ _ _ _ _ _ _ _ _ _ _27b. *P. villosissimum* var. *pseudopubescens*

27a. Panicum villosissimum Nash var. villosissimum *Fig. 156.*

Panicum nitidum var. *villosum* Gray, N. Am. Gram. & Cyp. 2:111. 1835.
Panicum dichotomum var. *villosum* (Gray) Vasey, Bull. U.S.D.A. Div. Bot. 8:31. 1889.
Culms spreading papillose-pilose; sheaths spreading papillose-pilose; ligule 4–5 mm long; upper surface of blades appressed long-pilose; panicle to 10 cm long; spikelets 2.0–2.3 mm long; 2n = 18 (Brown, 1948).

COMMON NAME: Hairy Panic Grass.
HABITAT: Sandy soil, often in woodlands.
RANGE: Massachusetts to Michigan and Kansas, south to Texas and Florida; Central America.
ILLINOIS DISTRIBUTION: Occasional throughout the state. This taxon differs from var. *pseudopubescens* in its horizontally spreading pubescence and its longer ligules. The similar *P. scoparioides* has stiffer pubescence on the vegetative parts, while *P. praecocius* has smaller spikelets.

27b. Panicum villosissimum Nash var. pseudopubescens (Nash) Fern. Rhodora 36:79. 1934. *Fig. 157.*

Panicum pseudopubescens Nash, Bull. Torrey Club 26:577. 1899.
Panicum euchlamydeum Shinners, Am. Midl. Nat. 32:170. 1944.
Panicum commonsianum var. *euchlamydeum* (Shinners) Pohl, Am. Midl. Nat. 38:507. 1947.
Culms appressed papillose-pilose; sheaths ascending papillose-pilose; ligule 2–3 mm long; upper surface of blades glabrous or short-pilose; panicle to 12 cm long; spikelets 2.2–2.4 mm long; 2n = 18 (Brown, 1948, as *P. pseudopubescens*).

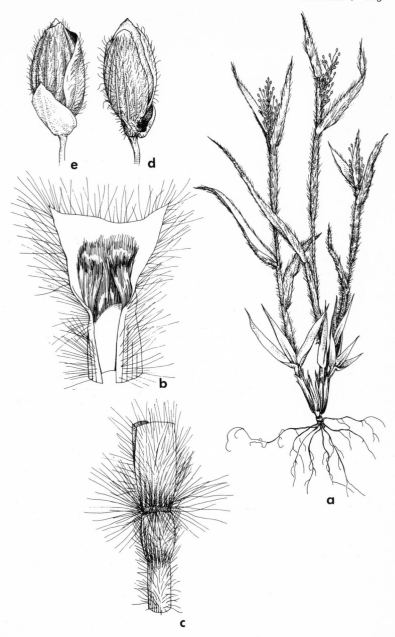

157. *Panicum villosissimum* var. *pseudopubescens* (Panic Grass). *a*. Habit, X½. *b*. Sheath, with ligule, X7½. *c*. Sheath and node, X7½. *d*. Spikelet, front view, X12½. *e*. Spikelet, back view, X12½.

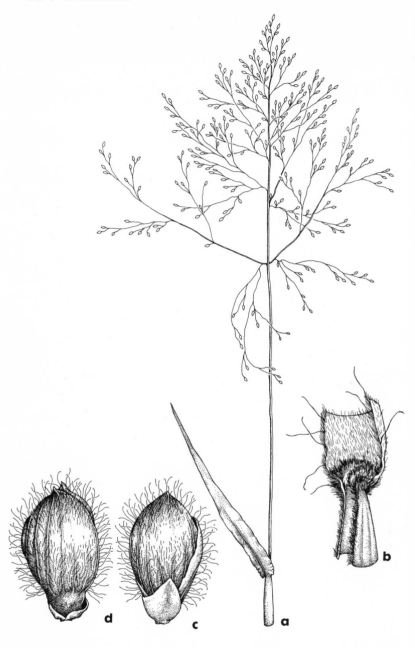

158. *Panicum scoparioides* (Panic Grass). *a.* Upper part of plant, X½. *b.* Sheath, with ligule, X5. *c.* Spikelet, front view, X17½. *d.* Spikelet, back view, X17½.

COMMON NAME: Panic Grass.

HABITAT: Sandy soil.

RANGE: Connecticut to Wisconsin and Kansas, south to Mississippi and Florida; Mexico.

ILLINOIS DISTRIBUTION: Occasional throughout the state; rare in the southern counties.

Panicum villosissimum var. *pseudopubescens* differs from var. *villosissimum* in its ascending pubescence of the culms and sheaths and in its shorter ligules.

28. Panicum scoparioides Ashe, Journ. Elisha Mitchell Sci. Soc. 15:53. 1898. *Fig. 158.*

Panicum villosissimum var. *scoparioides* (Ashe) Fern. Rhodora 36:79. 1934.

Tufted perennial; culms slender, erect to ascending, to 40 cm tall, light green, glabrous to sparsely papillose-hispid; sheaths glabrous to sparsely papillose-hispid; ligule 2–3 mm long; blades 6–10 mm broad, firm, ascending, sparsely hispid above, appressed-pubescent beneath, ciliate near the base; panicle to 7 cm long, nearly ⅔ as wide, densely flowered, the branches ascending; spikelets 2.2–2.3 mm long, 1.2 mm broad, obovoid, obtuse to subacute, papillose-pubescent; first glume ¼ as long as the spikelet, subacute, more or less puberulent; second glume and sterile lemma subequal, as long as the grain; grain 1.9 mm long, 1.1 mm broad, ellipsoid, obtuse to subacute.

COMMON NAME: Panic Grass.

HABITAT: Dry field.

RANGE: Vermont to Delaware; Michigan and Indiana to Minnesota and Kansas.

ILLINOIS DISTRIBUTION: Very rare; known only from Lake County (1½ miles northwest of Barrington, August 4, 1964, *Ozment 2871*).

This species has been considered a variety of *P. villosissimum,* but differs in its short, ascending pubescence of the culms and sheaths, the nearly glabrous upper surface of the blades, and the more densely flowered inflorescence. Fernald (1950) suggests it may be a hybrid between *P. oligosanthes* and a glabrous variety of *P. lanuginosum* (possibly var. *lindheimeri*).

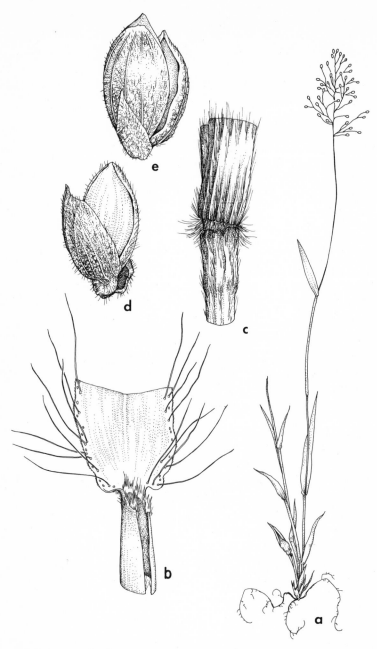

159. *Panicum columbianum* (Panic Grass). *a*. Habit, X½. *b*. Sheath, with ligule, X7½. *c*. Sheath and node, X7½. *d*. Spikelet, front view, X17½. *e*. Spikelet, back view, X17½.

SECTION *Columbiana*

29. Panicum columbianum Scribn. U.S.D.A. Div. Agrost. Bull. 7:78. 1897. *Fig. 159.*

Panicum tsugetorum Nash, Bull. Torrey Club 25:86. 1898.
Panicum lanuginosum siccanum Hitchc. & Chase, Rhodora 8:207. 1906.

Tufted perennial; culms spreading or ascending, to 60 cm tall, pale bluish-green, appressed-puberulent or ascending-pilose, with long hairs intermingled with short ones; sheaths pubescent with long and short hairs; ligule 1.0–1.5 mm long; blades to 8 mm broad, firm, ascending, the upper surface glabrous or long-hairy at base, rarely long-hairy throughout, the lower surface appressed-puberulent; panicle to 7.5 cm long, nearly as wide, the axis crisp-puberulent, the branches spreading; spikelets 1.8–1.9 mm long, 1.0 mm broad, obovoid, obtuse, short-pubescent; first glume ⅓–⅔ as long as the spikelet, acute, pubescent; second glume and sterile lemma equal, as long as the grain; grain 1.5 mm long, 1.0 mm broad, broadly ellipsoid, obtuse; autumnal form spreading, much branched, the blades scarcely reduced; 2n = 18 (Church, 1929).

COMMON NAME: Panic Grass.
HABITAT: Sandy woodlands.
RANGE: Maine to Wisconsin, south to Tennessee and Georgia.
ILLINOIS DISTRIBUTION: Rare; in the northern one-fourth of the state.

This is the only Illinois member of Section Columbiana, a group characterized by stiffly ascending blades, ligules up to 1.5 mm long, and much branched autumnal forms with scarcely reduced blades.

Our material, with spikelets 1.8–1.9 mm long, has been called *P. tsugetorum,* but no clear line of demarcation exists between this and typical *P. columbianum* which usually has spikelets 1.6–1.7 mm long.

SECTION *Sphaerocarpa*

30. Panicum sphaerocarpon Ell. Bot. S. C. & Ga. 1:125. 1816. *Fig. 160.*

Panicum dichotomum var. *sphaerocarpon* (Ell.) Wood, Class-book 786. 1861.

Panicum microcarpon var. *sphaerocarpon* (Ell.) Vasey, Grasses U. S. 12. 1883.

Tufted perennial; culms spreading, light green, to 60 cm tall, the nodes appressed-pubescent; sheaths glabrous, ciliate on the margins, occasionally with viscid spots between the nerves; ligule less than 1 mm long; blades to 17 mm broad, ascending, firm, scabrous above and on the ciliate margins, glabrous beneath; panicle to 12 cm long, nearly as wide, loosely flowered, the axis usually viscid-spotted, the branches ascending; spikelets 1.5–1.8 mm long, 1.0–1.3 mm broad, obovoid-spherical, obtuse, puberulent; first glume ¼ as long as the spikelet, obtuse, glabrous; second glume and sterile lemma equal, as long as the grain; grain 1.4–1.5 mm long, 1.0–1.2 mm broad, obovoid-spherical, obtuse; autumnal form mostly prostrate, sparsely branched, the blades and panicles scarcely reduced; 2n = 18 (Brown, 1948).

COMMON NAME: Panic Grass.

HABITAT: Sandy soil.

RANGE: Vermont to Michigan and Kansas, south to Texas and Florida; Mexico; Central America; Venezuela.

ILLINOIS DISTRIBUTION: Occasional in the southern two-fifths of the state; absent elsewhere.

This species differs from *P. polyanthes* in its pubescent nodes, its broader panicles, and its spreading culms. Both are the only Illinois representatives of Section Sphaerocarpa.

31. **Panicum polyanthes** Schult. Mantissa 2:257. 1824. *Fig. 161.*

Tufted perennial; culms stout, erect, light green, to 90 cm tall, the nodes more or less glabrous; sheaths glabrous except for the ciliate margins; ligule none, or up to 0.5 mm long; blades to 25 mm broad, ascending, rather thin, glabrous or scabrous above, glabrous below, scabrous and ciliate on the margins near the base; panicle to 25 cm long, ¼–½ as broad, densely flowered, the branches ascending; spikelets 1.5–1.6 mm long, 1.0–1.1 mm broad, obovoid-spherical, obtuse, puberulent; first glume ⅓–⅔ as long as the spikelet, obtuse or subacute; second glume and sterile lemma equal, as long as the grain; grain 1.4–1.5 mm long, 1.0–1.2

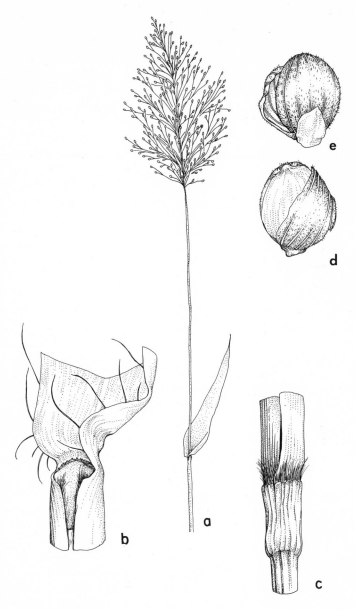

160. Panicum sphaerocarpon (Panic Grass). *a.* Inflorescence, X½. *b.* Sheath, with ligule, X7½. *c.* Sheath and node, X7½. *d.* Spikelet, front view, X17½. *e.* Spikelet, back view, X17½.

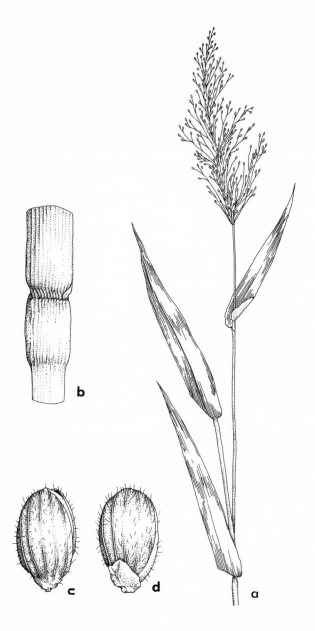

161. *Panicum polyanthes* (Panic Grass). *a.* Upper part of plant, X½. *b.* Sheath, with ligule, X7½. *c.* Sheath and node, X7½. *d.* Spikelet, front view, X17½. *e.* Spikelet, back view, X17½.

mm broad, obovoid-spherical, obtuse; autumnal form erect, sparsely branched, the blades and panicles reduced; 2n = 18 (Brown, 1948).

COMMON NAME: Panic Grass.
HABITAT: Low ground, chiefly in woodlands.
RANGE: Massachusetts to Illinois and Oklahoma, south to Texas and Georgia.
ILLINOIS DISTRIBUTION: Occasional in the southern one-third of the state; also in Peoria County; absent elsewhere.

SECTION **Oligosanthia**

32. **Panicum wilcoxianum** Vasey, Bull. U.S.D.A. Div. Bot. 8:32. 1889. *Fig. 162.*

Densely tufted perennial; culms erect, to 25 cm tall, papillose-hirsute, dull green, the nodes usually beardless; sheaths papillose-hirsute; ligule less than 1 mm long; blades to 6 mm broad, ascending to erect, firm, long-hirsute on both surfaces; panicle to 6 cm long, nearly ½ as broad, densely flowered, the branches spreading to ascending; spikelets 2.7–3.0 mm long, 1.5 mm broad, obovoid-ellipsoid, subacute, papillose-pubescent; first glume ⅓ as long as the spikelet, acute to obtuse, papillose-pubescent; second glume and sterile lemma subequal, slightly shorter than the grain; grain 2.4–2.5 mm long, 1.3–1.4 mm broad, ellipsoid, subacute; autumnal form much branched, the reduced blades exceeding the reduced panicles.

COMMON NAME: Panic Grass.
HABITAT: Dry soil, mostly in prairies.
RANGE: Alberta to Manitoba, south to New Mexico, Illinois, and Tennessee.
ILLINOIS DISTRIBUTION: Rare; limited to the extreme northwestern counties; also Pope County.
Section Oligosanthia, which includes *P. wilcoxianum,* *P. malacophyllum, P. oligosanthes, P. ravenelii,* and *P. leibergii,* is characterized by stout culms, small ligules (except *P. ravenelii*), and several-nerved spikelets ranging from 2.7–4.3 mm in length.

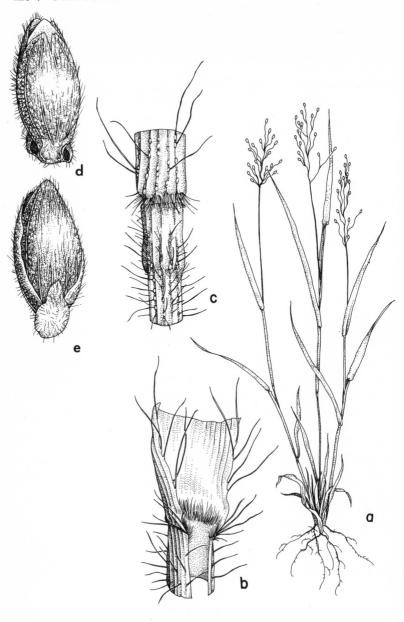

162. *Panicum wilcoxianum* (Panic Grass). *a.* Habit, X½. *b.* Sheath, with ligule, X7½. *c.* Sheath and node, X7½. *d.* Spikelet, front view, X12½. *e.* Spikelet, back view, X12½.

Panicum wilcoxianum differs from *P. malacophyllum* and *P. ravenelii* by the absence of velvety pubescence. It differs from *P. leibergii, P. oligosanthes* var. *oligosanthes,* and *P. oligosanthes* var. *scribnerianum* by its shorter spikelet (2.7–3.0 mm long in *P. wilcoxianum*). There is a strong similarity between *P. wilcoxianum* and *P. oligosanthes* var. *helleri,* but the latter taxon usually has broader blades and less pubescent stems and leaves.

Most specimens of *P. wilcoxianum* usually have beardless nodes, although the beard may be sparsely developed on occasion.

The Pope County specimens have been collected from the gravel knob areas in association with *Lechea villosa, Polygala incarnata,* and *Hypericum denticulatum.*

33. **Panicum malacophyllum** Nash, Bull. Torrey Club 24:198. 1897. *Fig. 163.*

Tufted perennial; culms erect, geniculate at the base, to 75 cm tall, velvety papillose-pilose, the nodes retrorsely bearded; sheaths papillose-pilose; ligule 1.0–1.5 mm long; blades to 12 mm broad, spreading or ascending, rather thin, velvety on both surfaces, ciliate toward the base; panicle to 7.5 cm long, less than ½ as broad, the branches spreading to strongly ascending; spikelets 2.9–3.0 mm long, 1.5–1.7 mm broad, ellipsoid-obovoid, subacute, papillose-pilose; first glume ⅓ as long as the spikelet, acute, papillose-pilose; second glume and sterile lemma equal, slightly shorter than the grain; grain 2.2 mm long, 1.5 mm broad, ellipsoid, subacute; autumnal form branched, sprawling, the blades little reduced.

COMMON NAME: Panic Grass.
HABITAT: Edge of limestone bluff; edge of dry woods.
RANGE: Illinois to Kansas, south to Texas and Tennessee.
ILLINOIS DISTRIBUTION: Rare; known only from Jackson County (Devil's Bake Oven, one mile north of Grand Tower, July 21, 1963, *J. Ozment 1071*) and Pope County (3½ miles southwest of Bay City, June 24, 1967, *J. Schwegman 1258*); also Johnson County. The velvety-pubescent culms and blades and the heavily bearded nodes relate this species to *P. ravenelii,* but these two species differ in spikelet size, with the spikelets averaging 3 mm long in *P. malacophyllum* and 4 mm long in *P. ravenelii.*

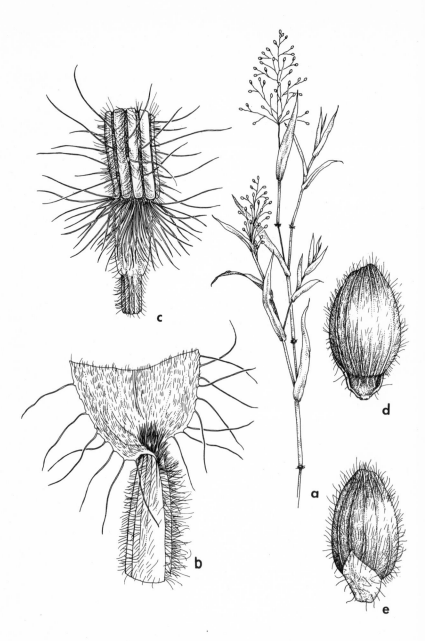

163. Panicum malacophyllum (Panic Grass). *a.* Upper part of plant, X½. *b.* Sheath, with ligule, X7½. *c.* Sheath and node, X7½. *d.* Spikelet, front view, X12½. *e.* Spikelet, back view, X12½.

Southern Illinois marks one of the limits of distribution for this grass.

34. **Panicum oligosanthes** Schult. Mantissa 2:256. 1824.

Panicum pauciflorum Ell. Bot. S.C. & Ga. 1:120. 1816, non R. Br. (1810).

Tufted perennial; culms spreading to ascending to erect, to 80 cm tall, glabrous, appressed-pubescent, harshly pubescent, or papillose-pilose; sheaths papillose-pubescent to glabrous; ligule 1–2 mm long; blades to 12 (–17) mm broad, spreading or ascending, glabrous or (rarely) sparsely long-hairy above, glabrous, puberulent, or harshly pubescent below, ciliate at the base; panicle to 15 cm long, nearly as broad, loosely flowered, the branches mostly ascending; spikelets 2.9–4.0 mm long, 1.6–2.0 mm broad, obovoid, obtuse to subacute, glabrous or sparsely pubescent; first glume ⅓–½ as long as the spikelet, acute, glabrous or sparsely pubescent; second glume and sterile lemma subequal, usually slightly shorter than the grain; grain 2.4–3.0 mm long, 1.5–1.9 mm broad, ellipsoid or ovoid, subacute to minutely apiculate; autumnal form spreading to ascending, much branched.

Three rather distinct varieties may be recognized in Illinois.

1. Spikelets 3.2–4.0 mm long; grains 2.8–3.0 mm long.
   2. Culms appressed-pubescent; first glume sparsely hirsute; grains 1.5–1.6 mm broad_____34a. *P. oligosanthes* var. *oligosanthes*
   2. Culms glabrous or spreading-pubescent; first glume glabrous; grains 1.8–1.9 mm broad_____
   _____34b. *P. oligosanthes* var. *scribnerianum*
1. Spikelets 2.9–3.0 mm long; grains 2.4–2.5 mm long_____
   _____34c. *P. oligosanthes* var. *helleri*

34a. **Panicum oligosanthes** Schult. var. **oligosanthes** *Fig. 164.*

Culms erect, to 80 cm tall, appressed-pubescent, olivaceous to purplish; sheaths appressed papillose-pubescent; ligule 1–2 mm long; blades to 10 mm broad, glabrous or (rarely) sparsely long-hairy above, harshly pubescent below; panicle to 12 cm long; spikelets 3.2–4.0 mm long, 1.7–1.9 mm broad, subacute, sparsely hirsute; first glume less than ½ as long as the spikelet, sparsely hirsute; grain 2.8–3.0 mm long, 1.5–1.6 mm broad, ellipsoid, subacute; 2n = 18 (Brown, 1948).

164. Panicum oligosanthes var. *oligosanthes* (Panic Grass). *a.* Upper part of plant, X½. *b.* Sheath, with ligule, X7½. *c.* Sheath and node, X7½. *d.* Spikelet, front view, X10. *e.* Spikelet, back view, X10.

COMMON NAME: Panic Grass.

HABITAT: Sandy soil, mostly in woodlands.

RANGE: Massachusetts to Michigan and Iowa, south to Texas and Florida.

ILLINOIS DISTRIBUTION: Occasional throughout the state. Specimens with shorter spikelets (3.2–3.4 mm long) are often very difficult to distinguish from var. *scribnerianum*. The type of pubescence which is used to differentiate these two taxa is not always clear-cut.

34b. Panicum oligosanthes Schult. var. **scribnerianum** (Nash) Fern. Rhodora 36:80. 1934. *Fig. 165.*

Panicum scribnerianum Nash, Bull. Torrey Club 22:421. 1895. Culms erect or ascending, geniculate at the base, to 50 cm tall, glabrous, harshly pubescent, or papillose-pilose; sheaths papillose-hispid to appressed-pubescent to nearly glabrous, ciliate on the margins; ligule about 1 mm long; blades to 12 (–17) mm broad, the upper surface glabrous, the lower surface appressed-pubescent to glabrous; panicle to 10 cm long; spikelets 3.2–3.3 mm long, 2 mm broad, obtuse, sparsely puberulent to nearly glabrous; first glume ⅓ as long as the spikelet, glabrous; grain 2.8–2.9 mm long, 1.8–1.9 mm broad, broadly ellipsoid, minutely apiculate; 2n = 18 (Church, 1929, as *P. scribnerianum*).

COMMON NAME: Panic Grass.

HABITAT: Dry, sandy soil, often in prairies.

RANGE: Maine to British Columbia, south to California, Mississippi, and Virginia; Mexico.

ILLINOIS DISTRIBUTION: Occasional throughout the state. Intergradation between this taxon and var. *oligosanthes* does not permit specific segregation of the two.

34c. Panicum oligosanthes Schult. var. **helleri** (Nash) Fern. Rhodora 36:80. 1934. *Fig. 166.*

Panicum helleri Nash, Bull. Torrey Club 26:572. 1899. Culms slender, spreading to ascending, to 60 cm tall, light bluish-green, appressed-pilose below, glabrous above; sheaths papillose-hispid to glabrous, ciliate on the margins; ligule about 1 mm long;

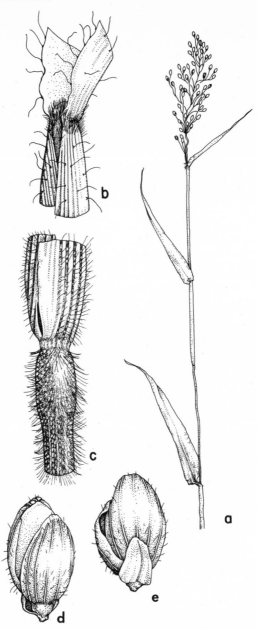

165. *Panicum oligosanthes* var. *scribnerianum* (Panic Grass). *a.* Upper part of plant, X½. *b.* Sheath, with ligule, X7½. *c.* Sheath and node, X7½. *d.* Spikelet, front view, X10. *e.* Spikelet, back view, X10.

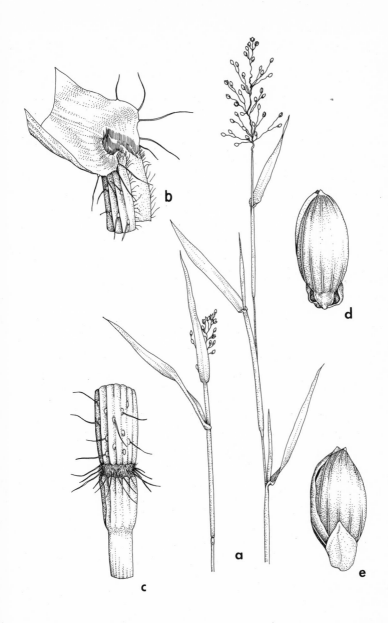

166. *Panicum oligosanthes* var. *helleri* (Panic Grass). *a.* Upper part of plants, X½. *b.* Sheath, with ligule, X7½. *c.* Sheath and node, X7½. *d.* Spikelet, front view, X12½. *e.* Spikelet, back view, X12½.

167. Panicum ravenelii (Ravenel's Panic Grass). *a.* Upper part of plant, X½. *b.* Sheath, with ligule, X2½. *c.* Spikelet, front view, X10. *d.* Spikelet, back view, X10.

blades to 12 mm broad, glabrous above, glabrous or puberulent below; panicle to 15 cm long; spikelets 2.9–3.0 mm long, 1.6–1.7 mm broad, obtuse, glabrous or sparsely pubescent; first glume ⅓ as long as the spikelet, glabrous; grain 2.4–2.5 mm long, 1.5–1.6 mm broad, ovoid, obscurely apiculate.

COMMON NAME: Panic Grass.
HABITAT: Exposed limestone ledge (in Illinois).
RANGE: Illinois to New Mexico and Louisiana.
ILLINOIS DISTRIBUTION: Very rare; known only from Randolph County (two miles north of Prairie du Rocher, May 25, 1962, *J. Ozment 791*).
Panicum oligosanthes var. *helleri* differs from the other varieties of *P. oligosanthes* by its smaller spikelets.
The Illinois collection cited above, along with a specimen from St. Louis County, Missouri, mark the northeast limit of distribution for this southern taxon.

35. Panicum ravenelii Scribn. & Merr. Bull. U.S.D.A. Div. Agrost. 24:36. 1901. *Fig. 167.*

Tufted perennial; culms rather stout, erect, to 75 cm tall, appressed papillose-pubescent, the nodes bearded; sheaths appressed papillose-pubescent; ligule 3–4 mm long; blades to 20 mm broad, more or less glabrous above, densely velvety-pubescent beneath, rounded at the base; panicle to 12 cm long, nearly as broad, shortly exserted, the branches mostly ascending, few-flowered; spikelets 4.0–4.3 mm long, 2.0–2.2 mm broad, ellipsoid, obtuse, sparsely papillose-pubescent; first glume about ⅓ as long as the spikelet, acute, sparsely pubescent; second glume and sterile lemma subequal, longer than the grain; grain averaging about 3.2 mm long, 2.0 mm broad, broadly ellipsoid, subacute; autumnal form spreading to ascending, much branched, with reduced leaves and panicles.

COMMON NAME: Ravenel's Panic Grass.
HABITAT: Cherty ravine and sandstone ledge (in Illinois).
RANGE: Delaware to southern Missouri and eastern Oklahoma, south to eastern Texas and northern Florida.
ILLINOIS DISTRIBUTION: Rare. (Hardin Co.: ¼ mile north of Lamb, *J. Schwegman 1941;* Union Co.: ½ mile south of Atwood Ridge Lookout Tower, *J. Schwegman 1916.*)
The stout plants, with broad, velvety leaves, retrorsely bearded nodes, and spikelets over 4 mm long, are distinctive from all other Illinois taxa of *Panicum* except *P. boscii* var. *molle,* a plant which is differentiated by its ex-

tremely short ligule. Both Illinois collections were made in July 1968.

36. Panicum leibergii (Vasey) Scribn. Britt. & Brown, Ill. Fl. 3:497. 1898. *Fig. 168.*

Panicum scoparium var. *leibergii* Vasey, Bull. U.S.D.A. Div. Bot. 8:32. 1889.

Panicum scribnerianum var. *leibergii* (Vasey) Scribn. U.S.D.A. Div. Agrost. 6:32. 1897.

Tufted perennial; culms slender, erect, geniculate at the base, to 75 cm tall, scabrous to pilose, dull green; sheaths papillose-hispid; ligule less than 0.5 mm long; blades to 15 mm broad, ascending or erect, rather thin, papillose-hispid on both surfaces, papillose-ciliate along the margins; panicle to 15 cm long, ⅓–½ as broad, the branches spreading to ascending; spikelets 3.7–4.0 mm long, 1.8–2.0 mm broad, oblong-obovoid, obtuse, papillose-hispid; first glume over ½ as long as the spikelet, acute, papillose-hispid; second glume and sterile lemma subequal, a little longer than the grain; grain 3 mm long, 1.7–1.8 mm broad, obovoid, obtuse to subacute; autumnal form leaning, sparsely branched, the blades little reduced.

COMMON NAME: Leiberg's Panic Grass.

HABITAT: Dry soil, mostly in sandstone woodlands or in prairies.

RANGE: New York to Manitoba, south to Kansas and Pennsylvania; Texas.

ILLINOIS DISTRIBUTION: Occasional in the northern half of the state; rare in the southern half.

This species, with its papillose hispidity throughout and with the rather large spikelets, is fairly distinct from all other species of *Panicum*. The ligule is highly reduced.

SECTION **Scoparia**

37. Panicum scoparium Lam. Encycl. 4:744. 1798. *Fig. 169.*

Perennial with a short knotty rhizome; culms stout, ascending, geniculate below, to over 1 m tall, gray-green, velvety, the nodes bearded but with a glabrous sticky ring immediately beneath them; sheaths velvety, except for the glabrous, sticky summit; blades to 18 mm broad, spreading to ascending, velvety-pubescent on both surfaces; panicle to 15 cm long, nearly as broad,

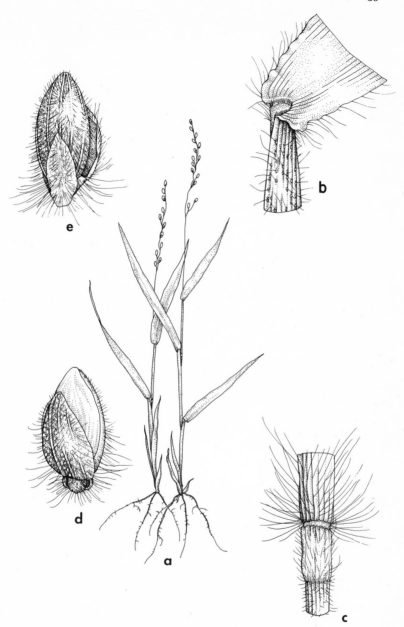

168. Panicum leibergii (Leiberg's Panic Grass). *a.* Habit, X½. *b.* Sheath, with ligule, X7½. *c.* Sheath and node, X7½. *d.* Spikelet, front view, X10. *e.* Spikelet, back view, X10.

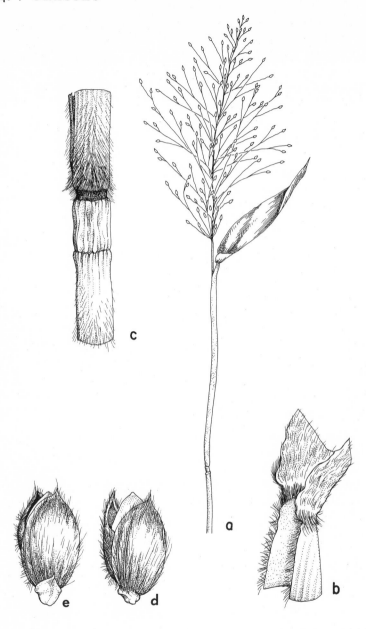

169. *Panicum scoparium* (Panic Grass). *a.* Upper part of plant, X½. *b.* Sheath, with ligule, X7½. *c.* Sheath and node, X7½. *d.* Spikelet, front view, X12½. *e.* Spikelet, back view, X12½.

pyramidal, the branches often with sticky patches; spikelets 2.4–2.6 mm long, 1.4–1.5 mm broad, obovoid, abruptly short-pointed at the tip, papillose-pubescent; first glume ¼ or less as long as the spikelet, acute, pubescent; second glume and sterile lemma subequal, not quite as long as the grain; grain 2.0–2.3 mm long, 1.4–1.5 mm broad, broadly ellipsoid, short-apiculate at the tip; autumnal form mostly spreading, much branched, with reduced leaves and panicles.

COMMON NAME: Panic Grass.

HABITAT: Moist fields and roadsides.

RANGE: Massachusetts to southern Missouri and eastern Oklahoma, south to Texas and Florida; Cuba.

ILLINOIS DISTRIBUTION: Known only from Pope County (3½ miles southwest of Bay City, June 27, 1967, *J. Schwegman 1260*).

Two additional stations in Pope County are known.

Despite the robust nature of this grass, *Panicum scoparium* was not discovered in Illinois until 1967. It is readily distinguished by the velvety pubescence of the culms and leaves and by the sticky ring immediately beneath each node on the culm.

SECTION **Commutata**

38. **Panicum commutatum** Schult. Mantissa 2:242. 1824.

Tufted perennial; culms erect, to 75 cm tall, glabrous to softly puberulent, usually purplish; sheaths glabrous to puberulent, short-ciliate; ligule up to 1 mm long; blades to 25 cm broad, spreading or ascending, firm, glabrous to puberulent, ciliate; panicle to 14 cm long, nearly as broad, loosely flowered, the branches spreading or ascending; spikelets 2.4–2.8 mm long, 1.2–1.3 mm broad, oblong-ellipsoid, obtuse to subacute, short-pubescent; first glume ¼–⅓ as long as the spikelet, obtuse to acute, glabrous to short-pubescent; second glume and sterile lemma subequal, nearly as long as the grain; grain 2.1–2.3 mm long, 1.1–1.2 mm broad, ellipsoid, minutely umbonate; autumnal form erect to reclining, branched, the blades somewhat reduced, the panicles greatly reduced.

Two varieties may be distinguished in Illinois.

1. Leaves to 25 mm broad, heart-shaped at base; spikelets 2.6–2.8 mm long_____38a. *P. commutatum* var. *commutatum*

1. Leaves to 10 mm broad, narrowed at base; spikelets 2.4–2.7 mm long_____38b. *P. commutatum* var. *ashei*

38a. Panicum commutatum Schult. var. commutatum *Fig. 170.*

Culms to 75 cm tall, glabrous or (rarely) softly puberulent, the nodes puberulent, often purplish; ligule less than 0.5 mm long; blades to 25 mm broad; panicle to 14 cm long; spikelets 2.6–2.8 mm long, 1.3 mm broad, obtuse; first glume ¼ as long as the spikelet, triangular, obtuse to acute, more or less glabrous; grain 2.2–2.3 mm long, 1.2 mm broad; autumnal form erect or nearly so, the blades somewhat reduced, the panicles greatly reduced; 2n = 18 (Brown, 1948).

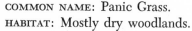

COMMON NAME: Panic Grass.

HABITAT: Mostly dry woodlands.

RANGE: Maine to Michigan and Oklahoma, south to Texas and Florida.

ILLINOIS DISTRIBUTION: Occasional in the southern one-third of the state; absent elsewhere.

This taxon is distinguished from var. *ashei* by its broader blades. It is distinguished from the broad-leaved *P. boscii, P. ravenelii,* and *P. latifolium* by its shorter spikelets, and from *P. clandestinum* by its lack of papillose pubescence.

38b. Panicum commutatum Schult. var. ashei Fern. Rhodora 36:83. 1934. *Fig. 171.*

Culms wiry, to 50 cm tall, puberulent; ligule less than 1 mm long; blades to 10 mm broad, more or less cordate at the base; panicle to 8 cm long; spikelets 2.4–2.7 mm long, 1.2–1.3 mm broad, obtuse to subacute; first glume ⅓ as long as the spikelet, subacute, short-pubescent; grain 2.1 mm long, 1.1 mm broad; autumnal form erect to reclining, the blades scarcely reduced; 2n = 18 (Brown, 1948, as *P. ashei*).

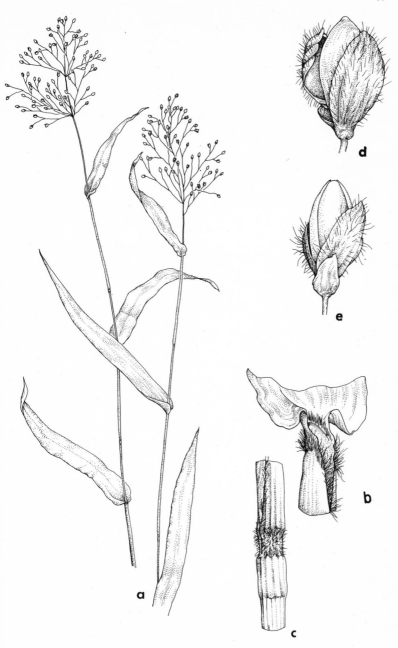

170. *Panicum commutatum* var. *commutatum* (Panic Grass). *a.* Upper part of plants, X½. *b.* Sheath, with ligule, X7½. *c.* Sheath and node, X7½. *d.* Spikelet, front view, X12½. *e.* Spikelet, back view, X12½.

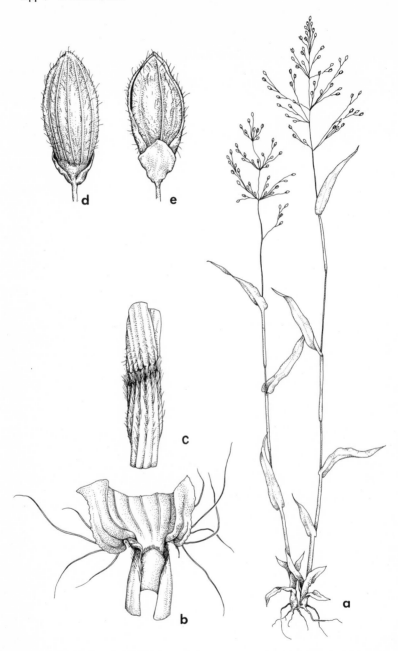

171. Panicum commutatum var. *ashei* (Panic Grass). *a.* Habit, X½. *b.* Sheath, with ligule, X7½. *c.* Sheath and node, X7½. *d.* Spikelet, front view, X12½. *e.* Spikelet, back view, X12½.

COMMON NAME: Panic Grass.

HABITAT: Dry woodlands.

RANGE: Massachusetts to Michigan and Missouri, south to Oklahoma and Florida.

ILLINOIS DISTRIBUTION: Rare; known only from Jackson County, where it was first collected in 1964.

Voss (1966) explains why this varietal name is not based on *Panicum ashei* Pearson in Ashe.

Although the width of the blades is strikingly different between var. *ashei* and var. *commutatum*, no other clear-cut differences can be found to merit specific segregation.

39. Panicum joori Vasey, U.S.D.A. Div. Bot. Bull. 8:31. 1889.

Fig. 172.

Panicum commutatum var. *joorii* (Vasey) Fern. Rhodora 39:388. 1937.

Tufted perennial; culms decumbent, to 50 cm tall, glabrous or nearly so, the lowest internodes purplish; sheaths glabrous, ciliate; ligule up to 1 mm long; blades to 1.5 cm broad, spreading to ascending, firm, glabrous except for some cilia near the base; panicle to 9 cm long, nearly as broad, loosely flowered, the branches spreading to ascending; spikelets 3.0–3.2 mm long, 1.2–1.4 mm broad, ellipsoid, abruptly short-pointed, pubescent; first glume about ⅓ as long as the spikelet, acute; second glume and sterile lemma papillose between the nerves, the sterile lemma conspicuously short-pointed; grain about 2.4 mm long, about 1.2 mm broad, ellipsoid, minutely umbonate; autumnal form not observed in Illinois but reported as widely spreading with branching from all the nodes and with reduced upper blades and numerous, small, partly included panicles.

COMMON NAME: Panic Grass.

HABITAT: Low, swampy woods.

RANGE: Virginia to Arkansas, south to southeastern Texas and Florida; southern Illinois; Mexico.

ILLINOIS DISTRIBUTION: Known only from a single collection (Johnson Co.: Heron Pond, August 4, 1969, *J. White 1042*).

Most recent authors have reduced *P. joori* to a variety of *P. commutatum,* and there is no question about a close relationship existing between these two entities. It ap-

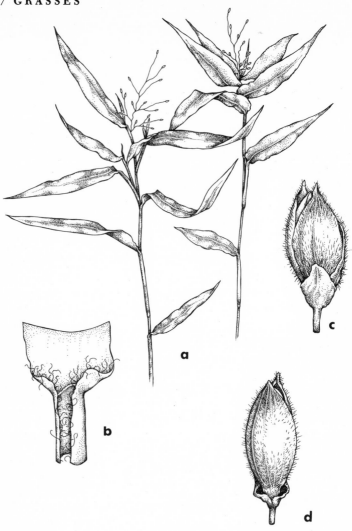

172. Panicum joori (Panic Grass). *a.* Upper part of plants, X½. *b.* Sheath, with ligule, X5. *c.* Spikelet, front view, X12½ *d.* Spikelet, back view, X12½.

pears to me, however, that the critical and distinctive pointed sterile lemma, coupled with a larger spikelet, decumbent culms, and generally narrower blades, is sufficient to maintain species status for *P. joori.*

The Illinois collection, made in 1969, is over 300 miles north of the nearest known station for *P. joori.*

SECTION **Latifolia**

40. Panicum clandestinum L. Sp. Pl. 58. 1753. *Fig. 173.*

Panicum latifolium L. var. *clandestinum* (L.) Pursh, Fl. Am. Sept. 1:68. 1814.

Densely tufted perennial from a stout rootstock; culms stout, erect, to 1.5 m tall, scabrous to papillose-hispid below the nodes; sheaths papillose-hispid to glabrous, with a puberulent ring at the summit; ligule 0.5 mm long; blades to 30 mm broad, spreading or reflexed, scabrous on both surfaces, ciliate near the base; panicle to 17 cm long, ¾ as broad, many-flowered, the flexuous branches ascending; spikelets 2.7–3.0 mm long, 1.4–1.5 mm broad, oblong-obovoid, subacute to acute, sparsely pubescent; first glume ⅓ as long as the spikelet, obtuse to subacute, glabrous or sparsely pubescent; second glume and sterile lemma subequal, a little shorter than the grain; grain 2.1–2.3 mm long, 1.2–1.3 mm broad, ellipsoid, acute; autumnal form more or less erect, sparsely branched, the panicles reduced; 2n = 36 (Brown 1948).

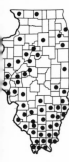

COMMON NAME: Broad-leaved Panic Grass.

HABITAT: Moist, sandy soil.

RANGE: Nova Scotia to Iowa, south to Texas and Florida.

ILLINOIS DISTRIBUTION: Common in the southern half of the state; less frequent northward.

This species, along with *P. commutatum, P. latifolium, P. boscii,* and *P. ravenelii,* has the broadest blades in the genus. It differs from *P. commutatum* in its papillose-pubescent sheaths, and from the other three in its shorter spikelets.

41. Panicum latifolium L. Sp. Pl. 58. 1753. *Fig. 174.*

Panicum schneckii Ashe, Bull. N. C. Agr. Exp. Sta. 175:116. 1900.

Tufted perennial from a knotted crown; culms rather stout, erect, to 1 m tall, glabrous or sparsely pubescent below, the nodes glabrous or rarely sparsely pubescent; sheaths ciliate along the margins, with a pubescent ring at the summit, otherwise glabrous; ligule less than 0.5 mm long; blades to 40 mm broad, spreading to ascending, rather thin, glabrous or (rarely) sparsely pubescent, short-ciliate at the base; panicle to 15 cm long, ⅔ as broad, few-flowered, the branches stiffly ascending; spikelets 3.4–3.7 mm long, 1.8–2.0 mm broad, ovoid to obovoid, obtuse to subacute,

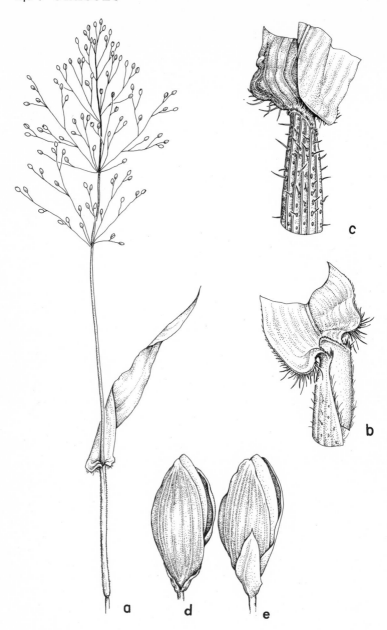

173. Panicum clandestinum (Broad-leaved Panic Grass). *a.* Upper part of plant, X½. *b.* Sheath, with ligule, X7½. *c.* Sheath and node, X7½. *d.* Spikelet, front view, X12½. *e.* Spikelet, back view, X12½.

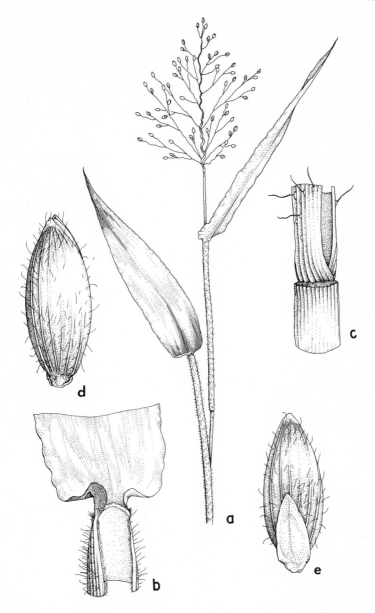

174. Panicum latifolium (Broad-leaved Panic Grass). *a.* Upper part of plant, X½. *b.* Sheath, with ligule, X7½. *c.* Sheath and node, X7½. *d.* Spikelet, front view, X12½. *e.* Spikelet, back view, X12½.

sparsely pubescent; first glume ⅓–½ as long as the spikelet, acute, glabrous to sparsely pubescent; second glume and sterile lemma equal, nearly as long as the grain; grain 3 mm long, 1.6–1.8 mm broad, ellipsoid, apiculate at the minutely pubescent apex; autumnal form spreading, branched.

COMMON NAME: Broad-leaved Panic Grass.
HABITAT: Dry, rocky woodlands.
RANGE: Maine to Minnesota, south to Arkansas and Georgia; Kansas.
ILLINOIS DISTRIBUTION: Occasional throughout the state. Although most specimens from Illinois are glabrous, a few may have pubescent culms and blades. This species has the broadest blades of any in the genus in Illinois, sometimes attaining a width of four centimeters.

Panicum schneckii, described from river bottoms in Illinois and Indiana, is identical with *P. latifolium.*

42. Panicum boscii Poir, in Lam. Encycl. Sup. 4:278. 1816.

Tufted perennial from a knotted crown; culms erect or ascending, to 75 cm tall, glabrous, puberulent, papillose-pilose, or downy-villous, the nodes retrorsely bearded; sheaths glabrous to downy-pilose, ciliate, with a puberulent ring at junction with blade; ligule about 1 mm long; blades to 30 mm broad, spreading, glabrous or appressed-pubescent above, glabrous to velvety below, ciliate at the base; panicle to 12 cm long, equally broad or broader, the axis puberulent to pilose, the branches spreading to ascending; spikelets 4.0–4.5 mm long, 2.0–2.2 mm broad, oblong-obovoid, papillose-pubescent; first glume ⅓–⅔ as long as the spikelet, acute, sparsely pubescent; second glume and sterile lemma unequal, the second glume shorter than the grain; grain 3.2–3.5 mm long, 1.5–1.6 mm broad, ellipsoid, minutely pubescent as the tip; autumnal form spreading, branched; 2n = 18, 36 (Brown, 1948).

Two varieties occur in Illinois.

1. Leaves glabrous or nearly so beneath; sheaths and culms glabrous or sparsely puberulent_____42a. *P. boscii* var. *boscii*
1. Leaves velvety-pubescent beneath; sheaths and culms softly pubescent_____42b. *P. boscii* var. *molle*

42a. Panicum boscii Poir. var. **boscii** *Fig. 175a–e.*

Vegetative parts glabrous or sparsely puberulent.

175. *Panicum boscii* (Large-fruited Panic Grass).—var. *boscii*. *a*. Upper part of plants, X½. *b*. Sheath, with ligule, X6. *c*. Sheath and node, X6. *d*. Spikelet, front view, X10. *e*. Spikelet, back view, X10.—var. *molle*. *f*. Sheath with node, X6.

COMMON NAME: Large-fruited Panic Grass.

HABITAT: Woodlands.

RANGE: Massachusetts to Wisconsin, south to Texas and Florida.

ILLINOIS DISTRIBUTION: Common in the southern half of the state; rare or absent elsewhere.

This taxon differs from *P. latifolium* in its larger spikelets and its retrorsely bearded nodes. It is one of the easiest kinds of *Panicum* to recognize in Illinois.

42b. Panicum boscii Poir. var. **molle** (Vasey) Hitchc. & Chase in Robinson, Rhodora 10:64. 1908. *Fig. 175f.*

Panicum latifolium L. var. *molle* Vasey ex Ward, Fl. Wash. 135. 1881.

Vegetative parts velvety or softly pubescent.

COMMON NAME: Large-fruited Panic Grass.

HABITAT: Woodlands.

RANGE: Massachusetts to Wisconsin, south to Texas and Florida.

ILLINOIS DISTRIBUTION: Occasional in the southern one-fourth of the state.

The similar *P. ravenelii* has longer ligules.

47. *Echinochloa* BEAUV. – Barnyard Grass

Annuals; blades flat; inflorescence paniculate, often contracted, composed of dense racemes; spikelets 1-flowered, arranged in 4 or more ranks; first glume up to half as long as the spikelet, 3-nerved; second glume and both lemmas equal in length, the second glume and sterile lemma usually awned, the fertile lemma papery, not awned.

Much variation exists in the treatment of North American species of *Echinochloa* by various authors. Wiegand's revision in 1921 is perhaps one of the more careful studies. Ali (1967) has also studied the genus.

KEY TO THE SPECIES OF *Echinochloa* IN ILLINOIS

1. Second glume with an awn 2–10 mm long; sheaths papillose-hirsute (glabrous in f. *laevigata*); grain about three times longer than broad_____1. *E. walteri*

1. Second glume awnless or with an awn less than 2 mm long; sheaths glabrous or scabrous; grain at most about twice as long as broad, usually shorter.
    2. Racemes of panicle slender, distant; grain 2.0–2.5 mm long; blades 2–6 mm broad_____2. *E. colonum*
    2. Racemes of panicle broader, more crowded; grain 2.5–3.5 mm long; blades 5–25 mm broad.
        3. Fertile lemma with a weak, easily broken tip, with a ring of minute setae just below the tip.
            4. Panicle green or purple; sterile lemma short-awned; pubescence of second glume and sterile lemma hispidulous or setose_____3. *E. crus-galli*
            4. Panicle dark brownish-purple; sterile lemma awnless; pubescence of second glume and sterile lemma appressed.___
            _____4. *E. frumentacea*
        3. Fertile lemma firm-tipped, without a setulose ring below the tip_____5. *E. pungens*

1. **Echinochloa walteri** (Pursh) Heller, Cat. N. Am. Pl. 2:21. 1900.
Panicum hirtellum Walt. Fl. Carol. 72. 1788, non L. (1759).
Panicum walteri Pursh, Fl. Am. Sept. 66. 1814.
Panicum hispidum Muhl. Descr. Gram. 107. 1817, non Forst. (1786).
Robust annual; culms erect, to 2 m tall; sheaths papillose-hirsute, rarely glabrous; blades harshly scabrous, 5–30 mm broad; panicle dense, 10–30 cm long, greenish, the racemes appressed-ascending; spikelets ellipsoid, 3.5–4.5 mm long; second glume strigose, with an awn 2–10 mm long; sterile lemma with an awn 10–25 mm long; fruit about three times longer than broad; anthers 0.6–1.2 mm long.

This is the only species of *Echinochloa* in Illinois with an awned second glume. The sterile lemma is longer awned than in any other species.

Two forms occur in Illinois.

1. Sheaths (at least the lower) papillose-hirsute_____
_____1a. *E. walteri* f. *walteri*
1. Sheaths glabrous_____1b. *E. walteri* f. *laevigata*

1a. **Echinochloa walteri** (Pursh) Heller f. **walteri** *Fig. 176a–c.*
Sheaths (at least the lower) papillose-hirsute.

HABITAT: Low ground; swamps; rarely in standing water. RANGE: Quebec to Minnesota, south to Texas and Florida; West Indies.

ILLINOIS DISTRIBUTION: Not common; scattered throughout the state.

This robust plant flowers from August to mid-October.

1b. Echinochloa walteri (Pursh) Heller f. laevigata Wieg.

Rhodora 23:62. 1921. *Fig. 176d.*

Panicum longisetum Torr. Am. Journ. Sci. 4:58. 1822.
Sheaths glabrous.

HABITAT: Same as f. *walteri.*
RANGE: Same as f. *walteri.*
ILLINOIS DISTRIBUTION: Known from Union County.

2. Echinochloa colonum (L.) Link, Hort. Berol. 2:209. 1833.

Fig. 177.

Panicum colonum L. Syst. Nat., ed. 10, 2:870. 1759.
Annual; culms often prostrate at the base, glabrous, to 80 cm tall; sheaths glabrous; blades glabrous, 2–6 mm broad; panicle slender, pale green, to 12 cm long; racemes distant, erect or ascending, to 3 cm long; spikelets crowded, arranged in 4 rows, obovoid or ovoid, 2–3 mm long; second glume and sterile lemma similar, appressed-pubescent or glabrous, awnless; grain 2.0–2.5 mm long, obtuse to acute; anthers 0.7–0.9 mm long.

*176. Echinochloa walteri.—*f. *walteri. a.* Inflorescence, X½. *b.* Sheath, with ligule, X7½. *c.* Spikelet, X7½—f. *laevigata. d.* Sheath, with ligule, X7½.

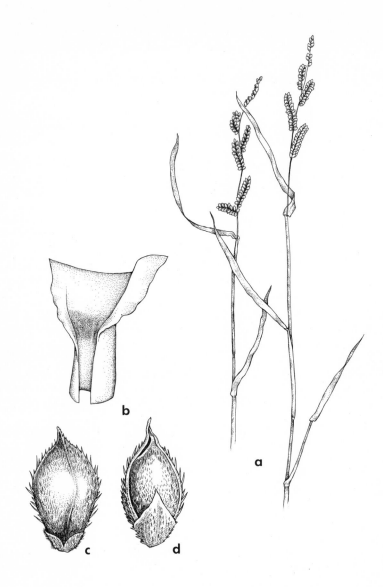

177. Echinochloa colonum (Jungle Rice). *a.* Upper part of plants, X½. *b.* Sheath, X7½. *c.* Spikelet, front view, X10. *d.* Spikelet, back view, X10.

COMMON NAME: Jungle Rice.

HABITAT: Waste ground.

RANGE: Native of Europe and Asia; occasionally found in the United States.

ILLINOIS DISTRIBUTION: Known only from a single collection from Cook County.

The slender, distant racemes and the nearly glabrous glumes and lemmas distinguish this species. The Illinois collection was made in mid-August.

3. **Echinochloa crus-galli** (L.) Beauv. Ess. Agrost. 53. 1812. *Fig. 178.*

Panicum crus-galli L. Sp. Pl. 56. 1753.

Echinochloa muricata var. *occidentalis* Wieg. Rhodora 23:58. 1921.

Echinochloa occidentalis (Wieg.) Rydb. Brittonia 1:82. 1931.

Echinochloa pungens var. *occidentalis* (Wieg.) Fern. & Grisc. Rhodora 37:137. 1935.

Annual; culms erect or decumbent from the base, usually less than 1 m tall; sheaths glabrous; blades 5–20 mm broad, glabrous or scabrous above; panicle to 25 cm long, green or purple, the mostly ascending racemes to 4 cm long; spikelets ovoid; second glume hispidulous on the nerves, acute; sterile lemma hispidulous on the nerves, rarely with marginal papillose hairs, short-awned; fertile lemma with a soft, easily broken tip, with a ring of small setae where the tip adjoins the body; grain 2.5–3.5 mm long, ovoid, shining.

COMMON NAME: Barnyard Grass.

HABITAT: Waste ground.

RANGE: Native of Europe and Asia; introduced throughout the United States.

ILLINOIS DISTRIBUTION: Occasional throughout the state. The ring of short setae where the tip of the fertile lemma adjoins the body is the best diagnostic character for this species. The treatment of the *Echinochloa crus-galli* complex in this work considers those specimens with firm-tipped fertile lemmas to belong to *E. pungens*. For the inclusion of *E. muricata* var. *occidentalis* within *E. crus-galli*, see Fassett (1949). The type of Wiegand's taxon is from Jackson County.

Barnyard Grass flowers during the summer months.

178. Echinochoa crus-galli (Barnyard Grass). *a.* Upper part of plant, X½. *b.* Sheath, with ligule, X2½. *c.* Spikelet, front view, X7½. *d.* Spikelet, back view, X7½.

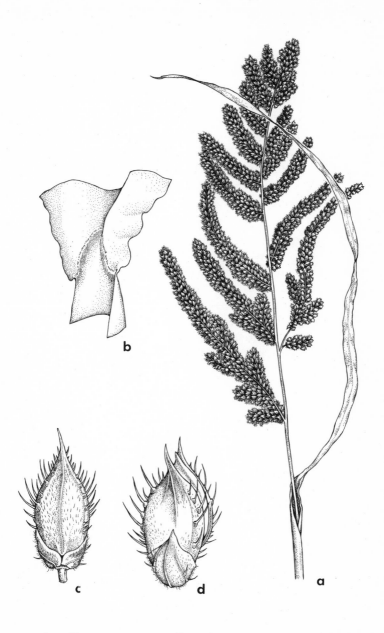

179. *Echinochloa frumentacea* (Billion Dollar Grass). *a.* Inflorescence, X½. *b.* Sheath, X7½. *c.* Spikelet, front view, X10. *d.* Spikelet, back view, X10.

4. **Echinochloa frumentacea** Link, Hort. Berol. 1:204. 1827.
Fig. 179.

Panicum frumentaceum Roxb. Fl. Ind. 1:307. 1820 non *P. frumentaceum* Salisb. (1796).
Echinochloa crus-galli var. *frumentacea* (Roxb.) W. F. Wight, Cent. Dict. Sup. 810. 1909.

Annual to 1.2 m tall; sheaths glabrous; blades 10–25 mm broad, glabrous; panicle dense, oblong-cylindric, to 15 cm long, brown-purple, the racemes arched-ascending; spikelets obtuse; second glume and sterile lemma glabrous or minutely appressed-pubescent, awnless; grain 2.5–3.5 mm long, ovoid; 2n = 36 (Hunter, 1934), 56 (Church, 1929a).

COMMON NAME: Billion Dollar Grass; Japanese Millet.
HABITAT: Waste ground.
RANGE: Native of Asia; spread from cultivation in the United States.
ILLINOIS DISTRIBUTION: Not common.

Some workers consider this taxon to be only a densely flowered variety of *E. crus-galli*. Both taxa do possess the setulose ring just beneath the tip of the fertile lemma.

5. **Echinochloa pungens** (Poir.) Rydb. Brittonia 1:81. 1931.

Panicum pungens Poir. Lam. Encycl. Sup. 4:273. 1816.

Annual; culms erect or decumbent, branching from the base; sheaths scabrous or glabrous; blades to 20 mm broad, scabrous above; panicle to 30 cm long, more or less open to contracted, purplish or greenish, the spreading to ascending racemes to 6 cm long; spikelets ovoid; second glume and sterile lemma papillose-hispid to minutely pubescent, the second glume acuminate to awn-tipped, the sterile lemma awnless or with an awn up to 10 mm long; grain 2.5–4.5 mm long; anthers 0.3–0.9 mm long.

Three rather distinctive varieties occur in Illinois.

1. Spikelets and grain 3.5–4.5 mm long; sterile lemma with an awn over 3 mm long; anthers 0.7–0.9 mm long; panicle more or less open, the branches spreading_____5a. *E. pungens* var. *pungens*
1. Spikelets and grain 2.5–3.5 mm long; sterile lemma awnless or with an awn less than 3 mm long; anthers 0.3–0.7 mm long; panicle contracted, the branches ascending.

2. Panicle purple; second glume and sterile lemma papillose-hispid
_____5b. *E. pungens* var. *microstachya*
2. Panicle green; second glume and sterile lemma appressed-
pubescent, not papillose_____5c. *E. pungens* var. *wiegandii*

5a. Echinochloa pungens (Poir.) Rydb. var. **pungens** *Fig. 180.*

Panicum muricatum Michx. Fl. Bor. Am. 1:47. 1803, non
Retz. (1786).

Setaria muricata Beauv. Ess. Agrost. 51. 1812.
Echinochloa muricata (Michx.) Fern. Rhodora 17:106. 1915.
Echinochloa crus-galli var. *muricata* (Michx.) Farw. Rep.
Mich. Acad. Sci. 21:350. 1920.

Panicle more or less open, to 30 cm long, purplish or greenish,
the racemes spreading; second glume acuminate or awn-tipped,
the sterile lemma with an awn over 3 mm long; grain 3.5–4.5 mm
long; anthers 0.7–0.9 mm long.

COMMON NAME: Barnyard Grass.
HABITAT: Low ground.
RANGE: Maine to Minnesota, south to Oklahoma and
North Carolina.
ILLINOIS DISTRIBUTION: Common throughout the state.
Although there is some intergradation between the three
taxa of *E. pungens*, var. *pungens* has longer spikelets
and grains, an open panicle, and longer anthers.
This taxon flowers from July to September.

5b. Echinochloa pungens (Poir.) Rydb. var. **microstachya**
(Wieg.) Fern. & Grisc. Rhodora 37:137. 1935. *Fig. 181.*

Echinochloa muricata var. *microstachya* Wieg. Rhodora 23:58.
1921.
Echinochloa microstachya (Wieg.) Rydb. Brittonia 1:82.
1931.

Panicle to 20 cm long, contracted, purple, the racemes ascending;
second glume and sterile lemma papillose-hispid, the second
glume awnless, the sterile lemma awnless or with an awn less
than 3 mm long; grain 2.5–3.5 mm long; anthers 0.3–0.7 mm long.

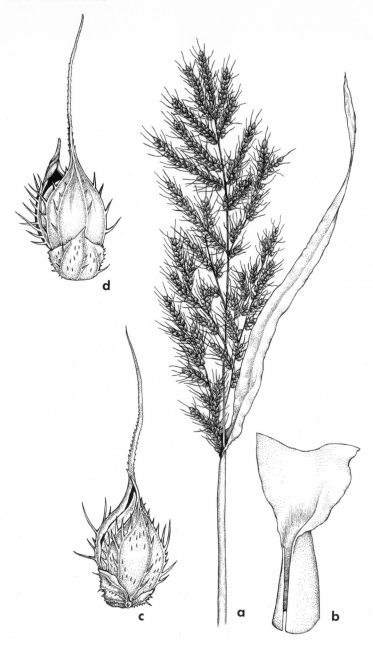

180. *Echinochloa pungens* var. *pungens* (Barnyard Grass). *a.* Inflorescence,
X½. *b.* Sheath, with ligule, X1½. *c.* Spikelet, front view, X7½. *d.* Spikelet,
back view, X7½.

181. Echinochloa pungens var. *microstachya* (Barnyard Grass). *a.* Inflorescence, X½. *b.* Sheath, with ligule, X1½. *c.* Spikelet, front view, X7½. *d.* Spikelet, back view, X7½.

182. *Echinochloa pungens* var. *wiegandii* (Barnyard Grass). *a.* Habit, X¼.
b. Sheath, with ligule, X4. *c.* Spikelet, front view, X10. *d.* Spikelet, back view, X10.

COMMON NAME: Barnyard Grass.

HABITAT: Low ground.

RANGE: Quebec to Wyoming, south to Texas and Pennsylvania; Mexico.

ILLINOIS DISTRIBUTION: Rare; known from Lake County. This taxon has the papillose-hispid glumes and sterile lemma of var. *pungens,* but has a smaller grain, smaller anthers, and a contracted panicle. *Echinochloa pungens* var. *microstachya* resembles *E. crus-galli,* but has a purple panicle and papillose-hispid glumes and sterile lemmas.

5c. Echinochloa pungens (Poir.) Rydb. var. **wiegandii** Fassett, Rhodora 51:2. 1949. *Fig. 182.*

Panicle contracted, to 30 cm long, green, the racemes ascending; second glume and sterile lemma minutely pubescent, not papillose, the second glume awnless, the sterile lemma awnless or with an awn less than 3 mm long; grain 2.5–3.5 mm long; anthers 0.3–0.7 mm long.

COMMON NAME: Barnyard Grass.

HABITAT: Low ground.

RANGE: Maine to Washington, south to California, Texas, and Rhode Island.

ILLINOIS DISTRIBUTION: Occasional throughout the state. A collection by Gleason from Jackson County, which served as the type for Wiegand's *E. muricata* var. *occidentalis,* is actually the introduced *E. crus-galli.* Although all other specimens cited as var. *occidentalis* are distinct, that epithet cannot be used. Fassett (1949) is the first to propose the epithet *wiegandii.*

48. *Setaria* BEAUV. – Foxtail Grass

Annuals or perennials; blades flat; inflorescence paniculate, contracted, spike-like; spikelets 1-flowered, each subtended by 1–20 scabrous bristles; first glume one-fourth to one-half as long as the spikelet, 3- to 5-nerved; fertile lemma rugose or rugulose, rarely smooth.

Setaria has been monographed by Rominger (1962).

KEY TO THE SPECIES OF Setaria IN ILLINOIS

1. Perennial from short, knotty rhizomes; second glume 1.0–1.5 mm long_____1. S. geniculata
1. Annuals from fibrous roots; second glume 1.8–2.7 mm long.
    2. Spike with retrorsely scabrous bristles_____2. S. verticillata
    2. Spike with antrorsely scabrous bristles.
        3. Each spikelet subtended by 5–20 bristles; blades more or less twisted_____3. S. lutescens
        3. Each spikelet subtended by 1–3 bristles; blades usually not twisted.
            4. Spike strongly arching; blades strigose above, puberulent below_____4. S. faberi
            4. Spike erect or scarcely arching; blades glabrous (sometimes ciliate).
                5. Spike usually not green, often lobed or interrupted; first glume short-acuminate, 1.2–1.5 mm long; fertile lemma smooth_____5. S. italica
                5. Spike usually green, unlobed (except var. *major*); first glume acute, 0.7–1.0 mm long; fertile lemma rugose or rugulose_____6. S. viridis

1. **Setaria geniculata** (Lam.) Beauv. Ess. Agrost. 51. 1812.
    Fig. 183.

Panicum geniculatum Lam. Encycl. 4:727. 1798.
Perennial from short, knotty rhizomes; culms compressed, geniculate at base, to 75 (–90) cm tall; blades flat, 2–8 mm broad; spike erect, to 10 cm long; spikelets 2–3 mm long, subtended by 4–12 bristles; bristles 2–12 mm long, antrorsely scabrous, purple or tawny; first glume 0.7–1.0 mm long, 3-nerved; second glume 1.0–1.5 mm long, 5-nerved; fertile lemma transversely rugose; 2n = 72 (Brown, 1948).

COMMON NAME: Perennial Foxtail.
HABITAT: Fields and roadsides.
RANGE: Massachusetts to California, east to Texas and Florida; Mexico; West Indies.
ILLINOIS DISTRIBUTION: Not common; apparently absent from the northern one-third of the state.
This is the only perennial species of *Setaria* in Illinois. Along with S. *lutescens*, it has more bristles per spikelet than the other species.

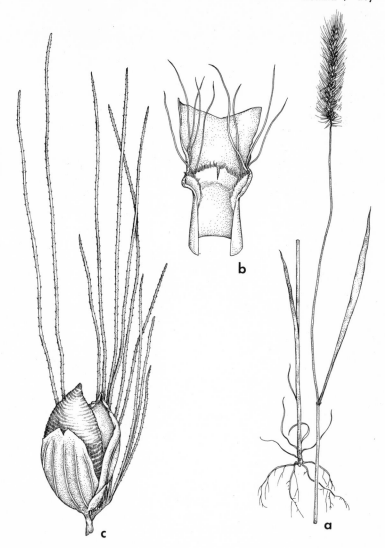

183. Setaria geniculata (Perennial Foxtail). *a.* Habit, X½. *b.* Sheath, with ligule, X7½. *c.* Spikelet, X12½.

Perennial Foxtail flowers from July to late September.

2. Setaria verticillata (L.) Beauv. Ess. Agrost. 51. 1812. *Fig. 184.*

Panicum verticillatum L. Sp. Pl., ed. 2, 82. 1762.

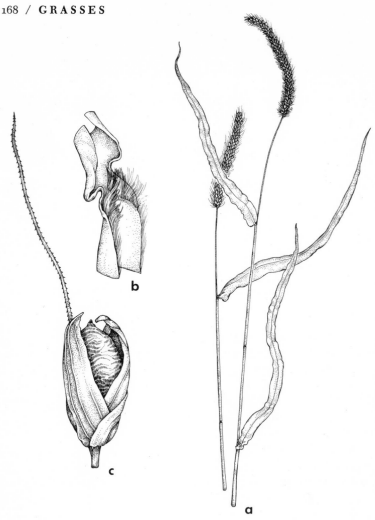

184. Setaria verticillata (Bristly Foxtail). *a.* Inflorescences, X½. *b.* Sheath, with ligule, X7½. *c.* Spikelet, X15.

Chaetochloa verticillata (L.) Scribn. Bull. U.S.D.A. Div. Agrost. 4:39. 1897.

Annual; culms branched from the base, more or less geniculate, to 75 (–90) cm tall; blades 5–15 mm broad, scabrous, more or less pilose, blue-green; spike to 15 cm long, erect; spikelets 2.0–2.2 mm long, each subtended by 1 bristle; bristle retrorsely

scabrous, purple or tawny, 4–7 mm long; first glume acute, 0.7–
1.0 mm long, 1- (3-) nerved; second glume subacute to obtuse,
1.8–2.2 mm long, 5-nerved; fertile lemma finely rugose; 2n = 36
(Rominger, 1962).

COMMON NAME: Bristly Foxtail.
HABITAT: Waste ground.
RANGE: Native of Europe and Asia; introduced throughout eastern United States.
ILLINOIS DISTRIBUTION: Occasional in the northern half
of the state; rare in the southern half.
The retrorsely scabrous bristles quickly permit identification of this species by merely stroking the spike between the thumb and index finger. This species flowers
from July to September.

3. **Setaria lutescens** (Weigel) Hubb. Rhodora 18:232. 1916.
Fig. 185.

Panicum lutescens Weigel, Obs. Bot. 20. 1772.
Setaria glauca Beauv. Ess. Agrost. 51:178. 1812, based on
Panicum glaucum L. (1753), which is actually *Pennisetum
glaucum* (L.) R. Br.
Annual; culms compressed, erect, branching at base, to 80 (–120)
cm tall; blades more or less twisted, 3–10 mm broad; spike to 15
cm long, erect; spikelets 3.0–3.5 mm long, subtended by 5–20
bristles; bristles 3–8 mm long, antrorsely scabrous, yellow or
tawny; first glume 1.5–1.8 mm long, acute, 3-nerved; second glume
2.0–2.5 mm long, 5- to 7-nerved; fertile lemma transversely rugose;
2n = 36 (Avdulov, 1931), 72 (Brown, 1948).

COMMON NAME: Yellow Foxtail.
HABITAT: Waste ground.
RANGE: Throughout North America.
ILLINOIS DISTRIBUTION: Common; in every county.
The transversely rugose fertile lemma and the spikelets
with five or more bristles subtending each relate this
species to the perennial S. *geniculata*. It is one of the
most common grasses in the state. It flowers from early
June to late September. Reeder (1951) and others have
argued that S. *glauca* should be the proper binomial for
this taxon.

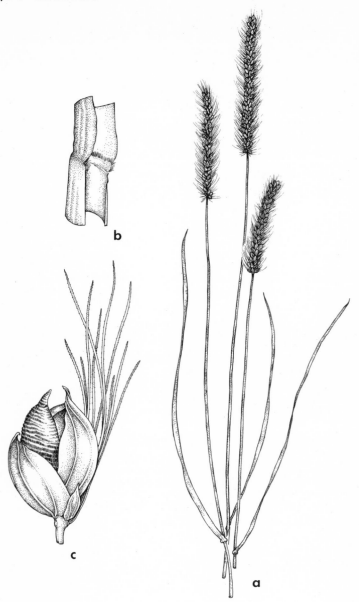

185. Setaria lutescens (Yellow Foxtail). *a.* Inflorescences, X½. *b.* Sheath, with ligule, X7½. *c.* Spikelet, X12½.

4. **Setaria faberi** Herrm. Beitr. Biol. Pflanz. 10:51. 1910. *Fig. 186.*

Annual; culms usually branched from the base, often geniculate, to nearly 2 m tall; blades to 17 mm broad, strigose above, puberulent below; spike strongly arching, to 17 cm long, to 3 cm thick; spikelets about 3 mm long, subtended by 1–3 bristles; bristles to 20 mm long, antrorsely scabrous, green or tawny; first glume 1 mm long, acute, 3-nerved; second glume 2.2 mm long, obtuse, 5- to 9-nerved; fertile lemma transversely wrinkled, 2.8 mm long; 2n = 18, 36 (Kishimoto, 1938).

COMMON NAME: Giant Foxtail.

HABITAT: Waste ground.

RANGE: Native of Asia; adventive throughout the eastern United States.

ILLINOIS DISTRIBUTION: Very common; in every county. The rapid spread of this species is unbelievable. It was described in 1910 from China, was first collected in Illinois in 1938, and is now a common weed in every county. Evers (1949) has described its rapid spread in Illinois.

5. **Setaria italica** (L.) Beauv. Ess. Agrost. 51. 1812. *Fig. 187.*

Panicum italicum L. Sp. Pl. 56. 1753.

Chaetochloa italica (L.) Scribn. Bull. U.S.D.A. Div. Agrost. 4:39. 1897.

Annual; culms erect, to nearly 1 m tall; blades 5–30 mm broad, glabrous; spike to 25 cm long, erect, sometimes lobed or interrupted; spikelets 2–3 mm long, subtended by 1–3 bristles; bristles to 10 mm long, antrorsely scabrous, purple, tawny, or (rarely) green; first glume short-acuminate, 1.2–1.5 mm long, 3-nerved; second glume acute, 2.0–2.7 mm long, 5- to 7-nerved; fertile lemma smooth; 2n = 18 (Avdulov, 1928).

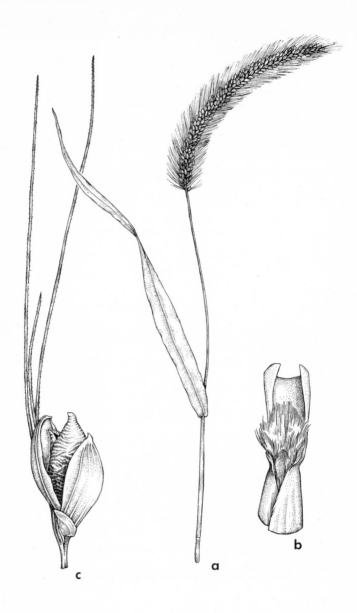

186. *Setaria faberi* (Giant Foxtail). *a*. Inflorescence, X½. *b*. Sheath with ligule, X7½. *c*. Spikelet, X12½.

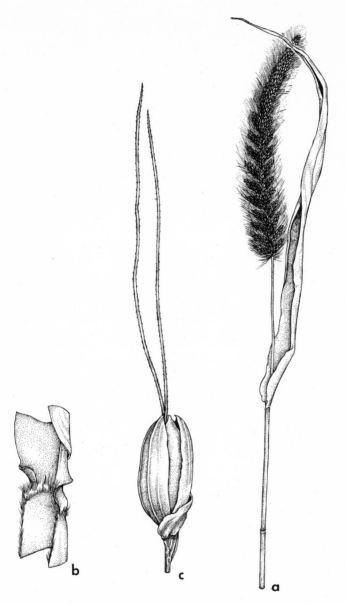

187. *Setaria italica* (Italian Millet). *a.* Inflorescence, X½. *b.* Sheath, with ligule, X7½. *c.* Spikelet, X12½.

COMMON NAME: Italian Millet; Hungarian Millet.

HABITAT: Waste ground.

RANGE: Native of Europe and Asia; introduced throughout the United States.

ILLINOIS DISTRIBUTION: Not common; so far collected only in the northern half of the state; also Williamson County. This species has a large spike like that of *S. faberi*, but the spike of *S. faberi* is strongly arching. *Setaria italica* differs from *S. viridis* in its more lobulate spikes, larger spikelets, and its longer, acuminate first glumes.

Italian Millet sometimes is cultivated in Illinois. It flowers from July to September.

6. Setaria viridis (L.) Beauv. Ess. Agrost. 51. 1812.

Panicum viride L. Syst. Nat., ed. 10, 870. 1759.

Annual; culms branched from the base, usually geniculate, to 75 cm (to 2.5 m in var. *major*) tall; blades to 12 (–25) mm broad, glabrous, often ciliate; spikes to 15 cm long, erect or a little arching; spikelets 1.8–2.2 mm long, subtended by 1–3 bristles; bristles to 10 mm long, antrorsely scabrous, green, purple, or tawny; first glume acute, 0.7–1.0 mm long, 3-nerved; second glume obtuse, 2.0–2.5 mm long, 5-nerved; fertile lemma finely rugose.

Two varieties occur in Illinois.

1. Culms to 75 cm tall; blades to 12 mm broad; panicle unlobed____
_____6a. *S. viridis* var. *viridis*
1. Culms 1.5–2.5 m tall; blades to 25 mm broad; panicle appearing lobed near base_____6b. *S. viridis* var. *major*

6a. Setaria viridis (L.) Beauv. var. **viridis** *Fig. 188a–c.*

Chaetochloa viridis (L.) Scribn. Bull. U.S.D.A. Div. Agrost. 4:39. 1897.

Culms to 75 cm tall; blades to 12 mm broad; panicle unlobed; 2n = 18 (Avdulov, 1931).

188. Setaria viridis (Common Foxtail).—var. *viridis.* *a.* Upper part of plants, X½. *b.* Sheath, with ligule, X7½. *c.* Spikelet, X15.—var. *major.* *d.* Inflorescence, X½.

COMMON NAME: Common Foxtail; Green Foxtail.
HABITAT: Waste ground.
RANGE: Native of Europe and Asia; introduced throughout North America.
ILLINOIS DISTRIBUTION: Common; in every county.
This very common, waste-ground grass flowers from June to September.

6b. Setaria viridis (L.) Beauv. var. **major** (Gaudin) Pospichal, Fl. Oesterr. Kustenl. 1:51. 1897. *Fig. 188d.*

Panicum viride maius Gaudin, Agrost. Helvet. 1:18. 1811.
Culms 1.5–2.5 m tall; blades to 25 mm broad; panicle appearing lobed near base; 2n = 18 (Rominger, 1962).

COMMON NAME: Giant Green Foxtail.
HABITAT: Cultivated fields.
RANGE: Native of Europe; introduced in the United States in Pennsylvania, Michigan, South Dakota, Nebraska, Iowa, and Illinois.
ILLINOIS DISTRIBUTION: Known from Champaign and LaSalle counties.

49. *Cenchrus* L. – Sand Bur

Annuals (in Illinois); blades flat; inflorescence a raceme of burs; spikelets 2-flowered, the lower staminate or sterile, in groups of 2–6, surrounded by a spiny bur composed of united bristles; glumes shorter than the lemma; fertile lemma somewhat indurate.

The most recent study on the taxonomy of the genus is by DeLisle (1963).

The united bristles represent sterile branchlets.

Only the following species occurs in Illinois.

1. Cenchrus longispinus (Hack.) Fern. Rhodora 45:388. 1943. *Fig. 189.*

Cenchrus echinatus f. *longispina* Hack. in Kneucker, Allg. Bot. Zeitschr. 9:169. 1903.
Annual; culms branches and geniculate at the base, to 90 cm tall; sheaths keeled, pilose on the margins and at the throat;

189. *Cenchrus longispinus* (Sand Bur). *a.* Upper part of plant, X½. *b.* Sheath, with ligule, X7½. *c.* Bur. X5. *d.* Spikelet, X7½.

blades to 7.2 mm broad, scabrous or sparsely pilose; raceme to 10.2 cm long; burs somewhat slender, retrorsely barbed, often purplish, to 7 mm long; spikelets sessile, 2–3 per bur, to 7.8 mm long; first glume 1.5–3.8 mm long, 1-nerved; second glume 4.4–6.0 mm long, 3- to 5-nerved; sterile lemma 5.0–6.5 mm long, 3- to 7-nerved; fertile lemma 5.8–7.6 mm long, 3-nerved; 2n = 34 (DeLisle, 1963).

COMMON NAME: Sand Bur.

HABITAT: Sandy soil.

RANGE: Massachusetts to Oregon, south to California, Texas, and Florida; Mexico; Central America; West Indies.

ILLINOIS DISTRIBUTION: Occasional throughout the state. No other species of grass in Illinois has the bur-like covering of the spikelets. *Cenchrus pauciflorus,* often the binomial applied to our species, actually refers to a different species. Sand Bur flowers from July to September.

Tribe *Andropogoneae*

Usually rather coarse perennials; spikelets paired, 1-flowered, or the flower frequently reduced; glumes (at least the first) hard, permanently enclosing the florets; lemmas often strongly reduced.

During disarticulation, usually both spikelets in a pair fall as a unit.

The treatment followed for tribe Andropogoneae includes genera previously segregated into the Tripsaceae. As a result, ten genera are recognized in this work as belonging to the Andropogoneae: *Miscanthus, Erianthus, Sorghum, Sorghastrum, Andropogon, Microstegium, Bothriochloa, Schizachyrium, Tripsacum,* and *Zea.*

50. *Miscanthus* ANDERSS. – Eulalia

Perennials; blades flat; inflorescence paniculate, contracted, composed of ascending racemes; spikelets 1-flowered, paired, uniform, the pedicels of the pair unequal; glumes subequal, hairy; sterile lemma shorter than the glumes; fertile lemma smaller than the sterile lemma, awned or awnless.

KEY TO THE SPECIES OF Miscanthus IN ILLINOIS

1. Fertile lemma awned; blades up to 10 mm broad_____1. *M. sinensis*
1. Fertile lemma awnless; blades 10–18 mm broad_____
_____2. *M. sacchariflorus*

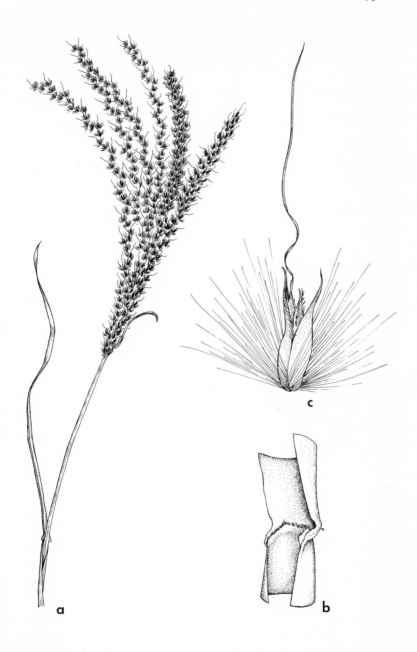

190. *Miscanthus sinensis* (Eulalia). *a.* Inflorescence, X½. *b.* Sheath, with ligule, X7½. *c.* Spikelet, X10.

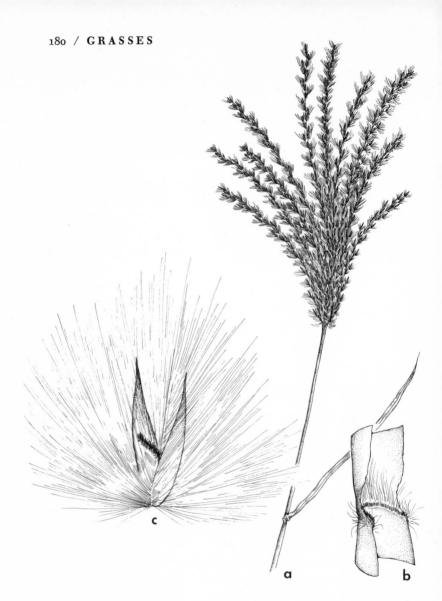

191. Miscanthus sacchariflorus (Plume Grass). *a.* Inflorescence, X½. *b.* Sheath, with ligule, X7½. *c.* Spikelet, X10.

1. **Miscanthus sinensis** Anderss. Ofv. Svensk. Vet. Akad. Forh. 12:166. 1856. *Fig. 190.*

Robust perennial with erect culms to 3 m tall; leaves mostly basal, to 10 mm broad, sharply serrate; panicle to 20 cm long, the numerous ascending racemes silky-plumose; spikelets with a tuft of hairs about as long as the glumes; fertile lemma awned; $2n = 42$ (Church, 1929).

COMMON NAME: Eulalia.

HABITAT: Along roads.

RANGE: Native of Asia; escaped from local cultivation in the United States.

ILLINOIS DISTRIBUTION: Collected thus far only from Jackson County.

This handsome grass flowers during September and October. The silky-plumose racemes make it particularly attractive.

2. **Miscanthus sacchariflorus** (Maxim.) Hack. in Engl. & Prantl, Pflanzenf. 2:23. 1887. *Fig. 191.*

Imperata sacchariflora Maxim. Mem. Acad. St. Petersb. Sav. Etrang. 9:331. 1859.

Robust perennial with erect culms to 4 m tall; leaves mostly basal, 10–18 mm broad; panicle to 40 cm long, the numerous ascending racemes very silky-plumose; spikelets with a tuft of hairs as long or longer than the glumes; fertile lemma awnless.

COMMON NAME: Plume Grass.

HABITAT: Waste ground.

RANGE: Native of Asia; rarely escaped in the United States.

ILLINOIS DISTRIBUTION: Known from three northern counties.

The awnless, fertile lemma, the broader blades, and the larger panicles distinguish this beautiful ornamental grass from the preceding.

51. *Erianthus* MICHX. – Plume Grass

Robust perennials; blades flat; inflorescence paniculate, composed of crowded racemes; spikelets borne in pairs, one sessile, the other pedicellate, uniform, each 1-flowered; glumes subequal, keeled; sterile lemma awnless; fertile lemma awned.

Mukherjee has revised the genus in 1958.

KEY TO THE SPECIES OF Erianthus IN ILLINOIS

1. Awn of fertile lemma up to 6 mm long; blades usually not more than 10 mm broad_____1. *E. ravennae*
1. Awn of fertile lemma 10–16 mm long; blades usually 10–25 mm broad.
    2. Panicle silvery, the axis villous; culms sericeous below the panicle; hairs subtending the spikelets much exceeding the spikelets; awn of fertile lemma twisted below_____2. *E. alopecuroides*
    2. Panicle tawny brown, the axis more or less glabrous; culms glabrous; hairs subtending the spikelets not exceeding the spikelets; awn of fertile lemma straight_____3. *E. brevibarbis*

1. **Erianthus ravennae** (L.) Beauv. Ess. Agrost. 14. 1812.

   Fig. 192.

   Andropogon ravennae L. Sp. Pl., ed. 2, 1481. 1763.
   Robust perennial to nearly 4 m tall; blades generally not over 10 mm broad; panicle erect, compact, the cream-colored racemes to 60 cm long; spikelets 4–6 mm long; awn of fertile lemma 3–6 mm long, straight.

COMMON NAME: Ravenna Grass.
HABITAT: Along roads (in Illinois).
RANGE: Native of Europe; escaped from local cultivation in the United States.
ILLINOIS DISTRIBUTION: Collected from a few counties in the southern half of Illinois.
This handsome ornamental, which has tendencies to escape from cultivation, flowers during August, September, and October. The short awns of the fertile lemma distinguish this species from the native species of *Erianthus* in Illinois. In addition, the inflorescence of *E. ravennae* is usually more compact.

2. **Erianthus alopecuroides** (L.) Ell. Bot. S.C. & Ga. 1:38. 1816. *Fig. 193*

   Andropogon alopecuroides L. Sp. Pl. 1045. 1753.
   Andropogon divaricatus L. Sp. Pl. 1045. 1753.
   Erianthus divaricatus (L.) Hitchc. Contr. U. S. Nat. Herb. 12:125. 1908.
   Robust perennial with culms to 3 m tall, sericeous below the

192. Erianthus ravennae (Ravenna Grass). *a.* Inflorescence, X½. *b.* Sheath, with ligule, X7½. *c.* Spikelet, X10.

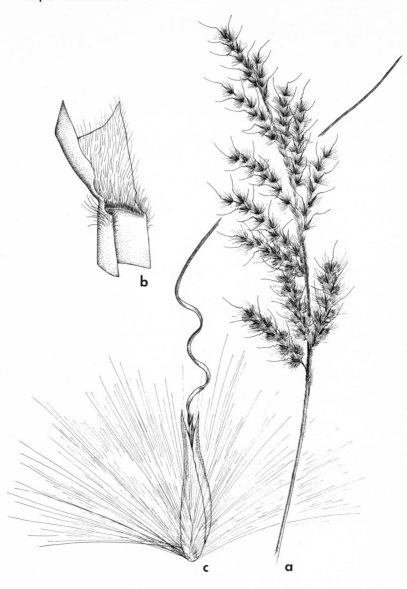

193. Erianthus alopecuroides (Plume Grass). *a.* Inflorescence, X½. *b.* Sheath, with ligule, X7½. *c.* Spikelet, X10.

panicle; sheaths pilose at the summit, otherwise glabrous; blades 10–25 mm broad, scabrous, densely pilose above near the base; panicle to 35 cm long, silvery to tawny, the axis villous; spikelets

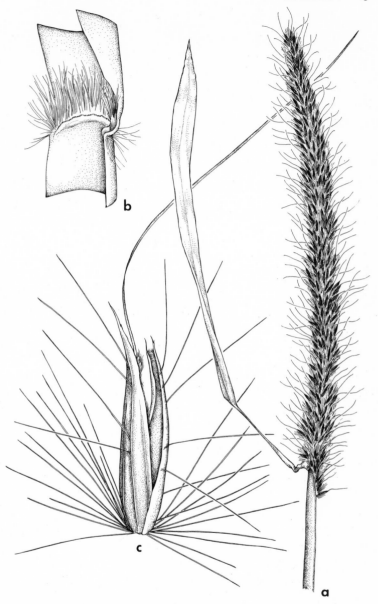

194. Erianthus brevibarbis (Brown Plume Grass). *a.* Inflorescence, X½. *b.* Sheath, with ligule, X7½. *c.* Spikelet, X10.

5–6 mm long, villous, subtended by hairs longer than the spikelets; awn of fertile lemma twisted below, 10–15 mm long.

COMMON NAME: Plume Grass.
HABITAT: Open woods.
RANGE: New Jersey to Oklahoma, south to Texas and Florida.
ILLINOIS DISTRIBUTION: Not common; restricted to the southern tip of the state.

This is one of the tallest native grasses in the state. Its handsome silvery or tawny panicle is borne during September and October. The twisted fertile awn is characteristic. There is evidence that this species is rapidly spreading throughout the southern counties of the state.

3. **Erianthus brevibarbis** Michx. Fl. Bor. Am. 1:55. 1803.
Fig. 194.

Perennial to 2 m tall; culms glabrous; sheaths glabrous or sparsely pubescent at the summit; blades 10–15 mm broad, scaberulous, pilose toward the base; panicle to 35 cm long, narrow, tawny brown, the axis more or less glabrous; spikelets 6–7 mm long, scabrous-puberulent toward the apex, subtended by hairs shorter than the spikelets; awn of fertile lemma not twisted, 15–16 mm long.

COMMON NAME: Brown Plume Grass.
HABITAT: Dry hills.
RANGE: Illinois and Arkansas.
ILLINOIS DISTRIBUTION: Only known from the type which is thought to have been collected somewhere in southwestern Illinois. This species is known elsewhere only from Pulaski County, Arkansas.
For an interesting discussion of this species and its original discovery, see Fernald (1945).

52. *Sorghum* MOENCH – Sorghum

Tall annuals or perennials; blades flat; inflorescence paniculate, terminal; spikelets borne in pairs, one sessile and perfect, the other pedicellate and staminate or sterile; glumes subequal, indurate; lemmas more or less indurate, usually awned.

The paniculate rather than racemose or spicate inflorescence distinguishes *Sorghum* from *Andropogon*. The lemmas in *Sorghum* tend to be more indurate than those in *Andropogon*.

KEY TO THE SPECIES OF Sorghum IN ILLINOIS

1. Perennials from rhizomes; awn of fertile lemma twisted near base.
   2. Culms up to 2 m tall; blades 10–20 mm broad; awn of fertile lemma 10–15 mm long_____1. S. halepense
   2. Culms more than 2 m tall; some or all of the blades 25–50 cm broad; awn of fertile lemma usually less than 10 mm long_____
   _____2. S. X almum
1. Annuals; awn of fertile lemma usually not twisted near base.
   3. Sessile spikelet 4.5–5.5 mm long, shorter than the pedicellate spikelet; awn of fertile lemma falling away early___3. S. bicolor
   3. Sessile spikelet 6–7 mm long, as long as the pedicellate spikelet; awn of fertile lemma persistent_____4. S. sudanense

1. Sorghum halepense (L.) Pers. Syn. Pl. 1:101. 1805. *Fig. 195.*

Holcus halepensis L. Sp. Pl. 1047. 1753.
Rhizomatous perennial; culms to nearly 2 m tall; blades 10–20 mm broad; panicle open, spreading, to 40 cm long; sessile spikelet 4.5–5.5 mm long; awn of fertile lemma 10–15 mm long, twisted near base, often falling away early; 2n = 40 (Huskins & Smith, 1934).

COMMON NAME: Johnson Grass.
HABITAT: Waste ground.
RANGE: Native of Europe and Asia; introduced in the eastern United States.
ILLINOIS DISTRIBUTION: Occasional to common in the southern half of the state, becoming less common northward.
In certain areas of southern Illinois, Johnson Grass has become an aggressive weed and is considered to be a primary noxious weed in Illinois. It is smaller in all respects than S. X *almum*.
Sorghum halepense flowers from June to October.

2. Sorghum X almum Parodi, Rev. Arg. Agron. 10:361–72. 1943. *Fig. 196.*

Rhizomatous perennial; culms more than 2 m tall; some or all of the blades 25–50 cm broad; panicle open, spreading, frequently well over 40 cm long; sessile spikelet 4.5–6.3 mm long; awn of

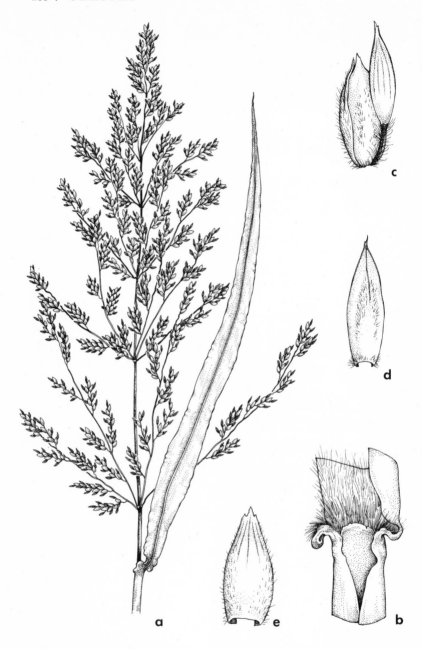

195. Sorghum halepense (Johnson Grass). *a.* Inflorescence, X½. *b.* Sheath, with ligule, X7½. *c.* Paired spikelets, X6. *d.* First glume, X7½. *e.* Second glume, X7½.

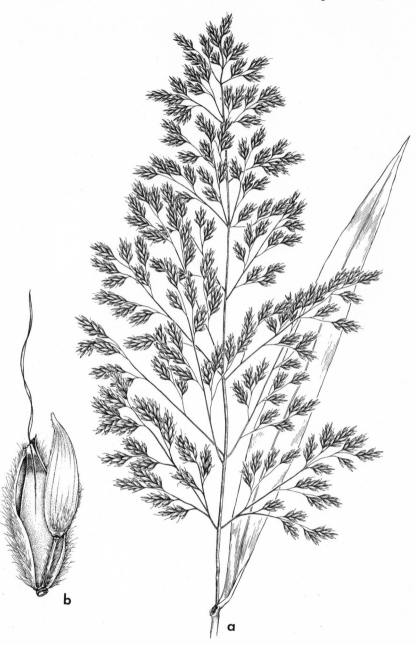

196. *Sorghum X almum* (Sorghum Grass). *a*. Inflorescence, X½. *b*. Paired spikelets, X6.

fertile lemma usually less than 10 mm long, twisted near base, often falling away early; 2n = 40 (Parodi, 1943).

COMMON NAME: Sorghum Grass.

HABITAT: Probably to be expected along roads and the edges of cultivated fields; scarcely established in Illinois at the time of this writing.

RANGE: Apparently originated in cultivation in Argentina.

ILLINOIS DISTRIBUTION: Thus far collected only in a few counties, but apparently more widespread.

Parodi (1943) considers Sorghum X almum to have originated under cultivation in Argentina as a hybrid between Johnson Grass and some other introduced sorghum.

Sorghum X *almum* resembles Johnson Grass in many ways, but is more robust and has coarser, larger stems, broader leaves, and longer panicles. The rhizomes do not appear to run as far or as deep as those of Johnson Grass. The seeds are virtually indistinguishable from those of Johnson Grass.

Early experiments on this grass in Illinois indicate that it will not overwinter in this state.

3a. **Sorghum bicolor** (L.) Moench, Meth. Pl. 207. 1794. *Fig. 197a–c.*

Holcus sorghum L. Sp. Pl. 1047. 1753.
Holcus bicolor L. Mant. Pl. 2:301. 1771.
Sorghum vulgare Pers. Syn. Pl. 1:101. 1805.

Annual; culms over 2 m tall; blades up to 50 mm broad; panicle open or narrow, to 75 cm long; sessile spikelet 4.5–5.5 mm long; awn of fertile lemma 10–15 mm long, usually falling away early; pedicellate spikelet longer than the sessile spikelet.

COMMON NAME: Sorghum.

HABITAT: Waste ground.

RANGE: Native of Asia; occasionally escaped from cultivation in the United States.

ILLINOIS DISTRIBUTION: Occasional throughout the state. Although this grass generally has been known as *Sorghum vulgare*, the epithet *bicolor* clearly has precedence over *vulgare*.

Several cultivated varieties of this species persist for only a few seasons in Illinois. The most adventive of

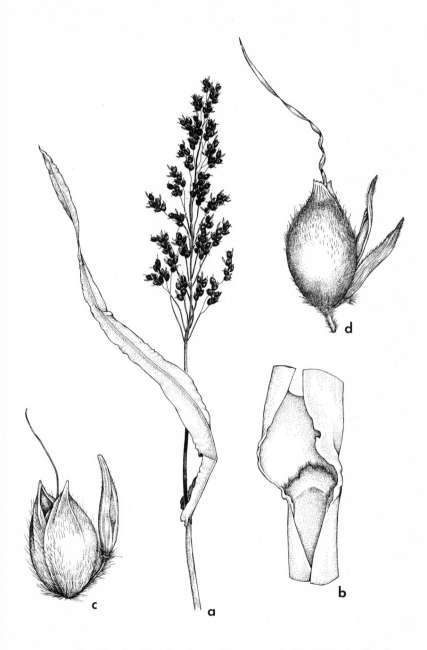

197. Sorghum bicolor (Sorghum). *a.* Upper tip of plant, X½. *b.* Sheath, with ligule, X5. *c.* Spikelet, X6.—var. *drummondii.* *d.* Spikelet, X6.

these is Chicken Corn (*S. bicolor* var. *drummondii*), a taxon up to 4 m tall, with blades nearly 5 cm broad and panicle very loose. It has been found along a railroad in Peoria County and in low cornfields in Gallatin and Massac counties. Two rarely escaped varieties are var. *caffrorum* (Kafir Corn) and var. *saccharatum* (Sugar Sorghum). Nomenclature for these varieties follows:

3b. **Sorghum bicolor** (L.) Moench var. **drummondii** (Nees) Mohlenbrock, comb. nov. *Fig. 197d.*

Andropogon drummondii Nees in Steud. Syn. Pl. Glum. 1:393. 1854.
Sorghum drummondii (Nees) Nees ex Hack. in DC. Mon. Phan. 6:507. 1889, in synon.
Sorghum vulgare Pers. var. *drummondii* (Nees) Hack. ex Chiov. Result. Sci. Miss. Stefan.-Paoli Somal. Ital. 1 Coll. Bot. 224. 1916.

3c. **Sorghum bicolor** (L.) Moench var. **caffrorum** (Retz.) Mohlenbrock, comb. nov.

Panicum caffrorum Retz. Obs Bot. 2:7. 1781.
Holcus caffrorum (Retz.) Thunb. Prodr. Pl. Cap. 1:20. 1794.
Sorghum caffrorum (Retz.) Beauv. Ess. Agrost. 131. 1812.
Sorghum vulgare Pers. var. *caffrorum* (Retz.) Hubb. & Rehder, Lflt. Harv. Univ. Bot. Mus. 1:10. 1932.

3d. **Sorghum bicolor** (L.) Moench var. **saccharatum** (L.) Mohlenbrock, comb. nov.

Holcus saccharatus L. Sp. Pl. 1047. 1753.
Sorghum saccharatum (L.) Moench, Meth. Pl. 207. 1794.
Sorghum vulgare Pers. var. *saccharatum* (L.) Boerl. Ann. Jard. Bot. Buitenz. 8:69. 1890.

4. **Sorghum sudanense** (Piper) Stapf in Prain, Fl. Trop. Afr. 9:113. 1917. *Fig. 198.*

Andropogon sorghum sudanensis Piper, Proc. Biol. Soc. Wash. 28:33. 1915.
Sorghum vulgare var. *sudanensis* (Piper) Hitchc. Jour. Wash. Acad. Sci. 17:147. 1927.

Annual; culms over 2 m tall; blades 8–12 mm broad; panicle to 30 cm long, erect, loose; sessile spikelet 6–7 mm long; awn of

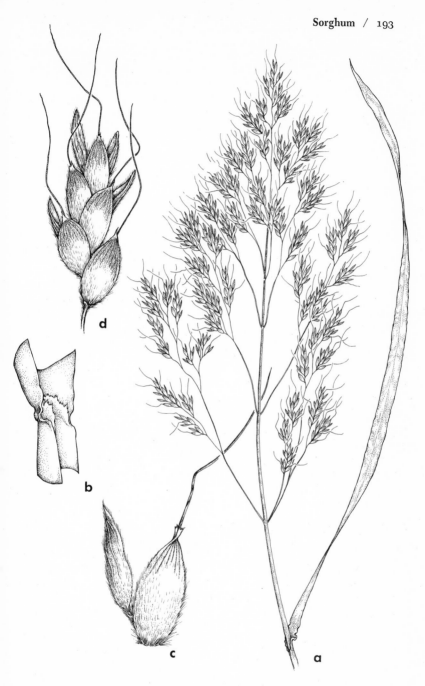

198. Sorghum sudanense (Sudan Grass). *a.* Inflorescence, X½. *b.* Sheath, with ligule, X2½. *c.* Paired spikelets, X7½. *d.* Cluster of spikelets, X5.

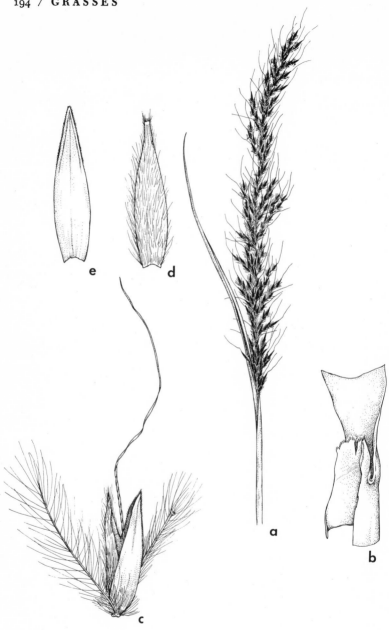

199. Sorghastrum nutans (Indian Grass). *a.* Inflorescence, X½. *b.* Sheath, with ligule, X5. *c.* Spikelet, X5. *d.* First glume, X6. *e.* Second glume, X6.

fertile lemma 10–15 mm long, persistent; pedicellate spikelet as long as the sessile spikelet; 2n = 20 (Huskins & Smith, 1934).

COMMON NAME: Sudan Grass.
HABITAT: Waste ground; introduced along a highway in Illinois.
RANGE: Native of Africa.
ILLINOIS DISTRIBUTION: Collected from Jackson County; undoubtedly in other parts of Illinois as well.

53. *Sorghastrum* NASH – Indian Grass

Tall perennials; blades flat; inflorescence paniculate, composed of racemes; spikelets borne in pairs, one sessile and perfect, the other composed only of a hairy pedicel; glumes of fertile spikelet subequal, indurate; fertile lemma thick, hyaline, awned.

Only the following species occurs in Illinois.

1. Sorghastrum nutans (L.) Nash in Small, Fl. Southeast.
U. S. 66. 1903. *Fig. 199.*
Andropogon nutans L. Sp. Pl. 1045. 1753.
Andropogon avenaceus Michx. Fl. Bor. Am. 1:58. 1803.
Andropogon ciliatus Ell. Bot. S.C. & Ga. 1:44. 1816.
Sorghum nutans (L.) Gray, Man. 617. 1848.
Chrysopogon nutans (L.) Gray, Man. 617. 1848.
Sorghastrum avenaceum (Michx.) Nash in Britton, Man. 71. 1901.
Perennial from scaly rhizomes; culms to over 2 m tall, sericeous at the nodes, otherwise glabrous; sheaths glabrous to hirsute; blades 5–10 mm broad, scabrous, glaucous; panicle to 35 cm long, narrow, the branchlets villous; spikelets lanceolate, 6–8 mm long; first glume villous; second glume glabrous or ciliate; awn of fertile lemma 10–15 mm long, bent near the base; sterile pedicel villous, 4–5 mm long; 2n = 40 (Brown, 1950).

COMMON NAME: Indian Grass.

HABITAT: Prairies, fields, open woodlands.

RANGE: Quebec to Wyoming, south to Texas and Florida; Mexico.

ILLINOIS DISTRIBUTION: Occasional to common throughout the state.

This species flowers from mid-August to early October. It differs from *Sorghum* in that the pedicel is all that represents the pedicellate spikelet.

Indian Grass is one of the characteristic tall grasses of the prairies. It may, however, be found in dry open woods.

The type for Michaux's *Andropogon avenaceus* is from Illinois.

54. Andropogon L. – Beardgrass

Perennials; blades usually flat; inflorescence spicate or racemose; spikelets borne in pairs, one sessile and perfect, the other pedicellate and staminate or sterile; glumes of fertile lemma membranous, usually awned.

The splitting up of *Andropogon* in Illinois into three genera follows the recommendation of Gould (1967). Although many readers will be disturbed by this arrangement, I can find no reason to reject Gould's treatment of the complex from a strictly taxonomic point of view.

Andropogon is retained for Big Bluestem (*A. gerardii*) and the Broom Sedges (*A. virginicus* and *A. elliottii*). Little Bluestem, formerly *A. scoparius,* is transferred to *Schizachyrium*, while Silver Beardgrass, formerly *A. saccharoides,* is placed in *Bothriochloa.*

KEY TO THE SPECIES OF Andropogon IN ILLINOIS

1. Pedicellate spikelet staminate, 3–10 mm long; upper sheaths not inflated; sessile spikelet 4.5–10.0 mm long_____1. *A. gerardii*
1. Pedicellate spikelet undeveloped, with only the villous pedicels present; upper sheaths inflated; sessile spikelet 2.5–5.0 mm long.
    2. Upper sheaths somewhat inflated, 2–6 cm long; culms more or less glabrous; racemes enclosed only at their base, the peduncles 2–10 mm long; awn straight or nearly so_____2. *A. virginicus*
    2. Upper sheaths greatly inflated, 6–12 mm long; culms villous at the upper nodes; racemes nearly entirely enclosed by the sheath, the peduncles more than 10 mm long; awn twisted or curved near base_____3. *A. elliottii*

1. **Andropogon gerardii** Vitman, Summa Pl. 6:16. 1792. *Fig.*
200.

Andropogon furcatus Muhl. in Willd. Sp. Pl. 4:919. 1806.
Large-tufted perennial from stout rhizomes; culms to 2 m tall;
sheaths glabrous to villous; blades 5–10 mm broad, scabrous,
often glaucous, the lower more or less villous; racemes 2–6, more
or less digitate, long-exserted, to 15 cm long, the rachis ciliate;
sessile spikelets 4.5–10.0 mm long; awn of fertile lemma 8–15
mm long, twisted below; pedicellate spikelet staminate, 3–10 mm
long, awnless.

COMMON NAME: Big Bluestem; Turkeyfoot Grass.
HABITAT: Prairies.
RANGE: Quebec to Saskatchewan, south to Arizona and
Florida.
ILLINOIS DISTRIBUTION: Occasional to common through-
out the state.
Big Bluestem flowers from July to late September. It is
the most robust of all species of *Andropogon* in Illinois,
and has the longest fertile spikelets. It is one of the most
characteristic species of prairies.

2. **Andropogon virginicus** L. Sp. Pl. 1046. 1753. *Fig. 201.*

Andropogon tetrastachyus Ell. Bot. S. D. & Ga. 1:150. 1816.
Andropogon virginicus var. *tetrastachyus* (Ell.) Hack. in DC.
Mon. Phan. 6:411. 1889.
Andropogon virginicus var. *genuinus* Fern. & Grisc. Rhodora
37:142. 1935.
Tufted perennial; culms more or less glabrous, to 1.5 m tall;
sheaths keeled, pilose; blades 3–6 mm broad, green or glaucous,
pilose on upper surface near the base; racemes 2–4, partly enclosed at the base in the sheath, 2.5–4.5 mm long, the peduncles
2–10 mm long, the rachis villous; upper sheaths somewhat inflated, 2–6 cm long; sessile spikelet 3–4 mm long; awn of fertile
lemma 10–20 mm long, nearly straight; sterile spikelet absent,
represented only by the villous pedicel; 2n = 20 (Church 1936),
40 (Saura, 1943).

200. *Andropogon gerardii* (Big Bluestem). *a*. Upper tip of plant, X½. *b*. Sheath, with ligule, X5. *c*. Paired spikelets, X7½.

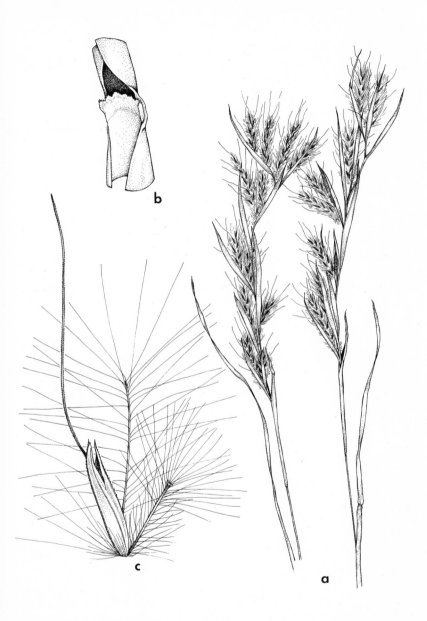

201. Andropogon virginicus (Broom Sedge). *a.* Upper tips of plants, X½. *b.* Sheath, with ligule, X5. *c.* Spikelet, X10.

COMMON NAME: Broom Sedge.

HABITAT: Fields and open woods.

RANGE: Massachusetts to Kansas, south to Texas and Florida.

ILLINOIS DISTRIBUTION: Occasional or common in the southern half of the state; absent in the northern half, except for Cook County.

Specimens with much branched inflorescences have been referred to var. *tetrastachyus,* but the exact demarcation between this variety and the typical variety does not exist.

Voigt (1953) has discussed the value of this species as a pasture grass.

Broom Sedge flowers from late August to mid-October.

3. **Andropogon elliottii** Chapm. Fl. South. U. S. 581. 1860.

Fig. 202.

Andropogon elliottii var. *projectus* Fern & Grisc. Rhodora 37:139. 1935.

Tufted perennial; culms to 90 cm tall, branched above, villous above, glabrous below; sheaths keeled, appressed-pubescent to nearly glabrous; blades 2.5–7.0 mm broad; racemes paired, often wholly enclosed in the sheath, 2–5 cm long, the peduncles over 10 mm long, the rachis villous; upper sheaths greatly inflated, coppery-colored, 6–12 cm long; sessile spikelet 3.5–5.0 mm long; awn of fertile lemma 10–25 mm long, twisted or curved near the base; sterile spikelet absent, represented only by the villous pedicel; $2n = 20$ (Hunter, 1934).

COMMON NAME: Elliott's Broom Sedge.

HABITAT: Fields, open woodlands.

RANGE: New Jersey to Missouri, south to Texas and Florida.

ILLINOIS DISTRIBUTION: Restricted to the southern tip of the state.

Voigt (1951) has made a detailed study of this species in Illinois.

This species, conspicuous because of its inflated, coppery-colored upper sheaths, flowers during September and October.

202. Andropogon elliottii (Elliott's Broom Sedge). *a.* Upper tip of plant, X½. *b.* Sheath, with ligule, X5. *c.* Spikelet, X7½. *d.* Inflated sheath, X1½.

55. *Microstegium* NEES

Weak, decumbent annuals; inflorescence racemose; spikelets uniform, paired, with one sessile and one pedicellate; glumes equal, not indurate; fertile lemma awned or awnless.

Only the following species occurs in Illinois.

1. **Microstegium vimineum** (Trin.) A. Camus, Ann. Soc. Linn.

Lyon 68:201. 1921. *Fig. 203.*

Andropogon vimineus Trin. Mem. Acad. St. Petersb. VI. Math. Phys. Nat. 2:268. 1832.
Eulalia viminea (Trin.) Kuntze, Rev. Gen. Pl. 2:775. 1891.
Pollinia viminea (Trin.) Merr. Enum. Philipp. Pl. 1:35. 1922.
Branched, reclining annual rooting from the lower nodes, the culms at length ascending near the tips; blades to 1 cm broad, narrowed at base to a short stalk; racemes 1–6, exserted, to 6 cm long; spikelets 4.5–6.0 mm long; second glume conspicuously ciliate on the keel; fertile lemma awnless.

COMMON NAME: Eulalia.

HABITAT: Moist, waste ground (in Illinois).

RANGE: Native to Asia; adventive in a few of the eastern states.

ILLINOIS DISTRIBUTION: Massac Co.: Fort Massac State Park, October 28, 1967, *J. Schwegman 1503.* Pope Co.: Two miles south of Bay City, September 21, 1967, *J. Schwegman 1454.*

This sprawling annual is becoming established at several areas along the Ohio River in Pope and Massac counties. The uniform paired spikelets in which one is sessile and one pedicellate are characteristic for this species. Our species at one time was placed in the genus *Eulalia.*

56. *Bothriochloa* KUNTZE

Perennials; blades flat; inflorescence paniculate; spikelets paired, the sessile one fertile and awned, the pedicellate one sterile or staminate.

Under traditional systems of grass classification, this genus was treated as a section (Amphilophis) of *Andropogon.*

Only one species occurs in Illinois.

203. *Microstegium vimineum* (Eulalia). *a.* Upper part of plants, X½. *b.* Sheath, with ligule, X5. *c.* Spikelet, X7½. *d.* Glume, X7½.

1. Bothriochloa saccharoides (Swartz) Rydb. Brittonia 1:81. 1931. *Fig. 204.*

Andropogon saccharoides Swartz, Prod. Veg. Ind. Occ. 26. 1788.

Amphilophis saccharoides (Swartz) Nash, N. Am. Fl. 17:125. 1912.

204. *Bothriochloa saccharoides* (Silver Beardgrass). *a.* Inflorescence, X½. *b.* Sheath, with ligule, X7½. *c.* Spikelet, X7½.

Tufted perennial; culms to about 1 m or more tall, glabrous to appressed-pubescent; blades flat, more or less glaucous, glabrous or nearly so, to 8 mm broad; panicle exserted, to 15 cm long, silvery-silky, composed of several elongated racemes to 4 cm long, the joints of the villous rachis strongly flattened; sessile spikelet fertile, 3–4 mm long, the lemma with a twisted awn to 15 mm long; pedicellate spikelet sterile, reduced to a single scale.

COMMON NAME: Silver Beardgrass.

HABITAT: Vacant field (in Illinois).

RANGE: Missouri and Alabama, westward to southern California; Mexico; apparently adventive in Illinois.

ILLINOIS DISTRIBUTION: Known only from Alexander (Thebes, October 4, 1966, *G. S. Winterringer 23679*) and Sangamon (along Illinois Central Railroad N of Glenarm, September 19, 1969, *J. White 1900*) counties. This handsome grass, native primarily to the southwestern United States, occurs in a vacant field in the middle of the community of Thebes where it grows with several other weedy species. The station is not more than one-quarter mile from the Mississippi River.

The large, silver inflorescence is very distinctive, and is mature by September.

This species in the past has usually been included in *Andropogon.*

57. *Schizachyrium* NEES

Rhizomatous perennials; blades flat or folded; culms much branched, each branch terminated by a single raceme; spikelets paired, the sessile one fertile and awned, the pedicellate one sterile.

This segregate genus of *Andropogon* is represented in Illinois by a single species.

1. Schizachyrium scoparium (Michx.) Nash in Small, Fl. S. E.

U. S. 59. 1903. *Fig. 205.*

Andropogon scoparius Michx. Fl. Bor. Am. 1:57. 1803.

Andropogon purpurascens Muhl. in Willd. Sp. Pl. 4:913. 1806.

Andropogon neo-mexicanus Nash, Bull. Torrey Club 25:83. 1898.

Andropogon scoparius var. *polycladus* Scribn. & Ball, Bull. U.S.D.A. Div. Agrost. 24:40. 1901.

Andropogon scoparius var. *villosissimus* Kearney in Scribn. & Ball, Bull. U.S.D.A. Div. Agrost. 24:41. 1901.

Andropogon scoparius var. *frequens* Hubb. Rhodora 19:103. 1917.

Andropogon scoparius var. *neomexicanus* (Nash) Hitchc. Proc. Biol. Soc. Wash. 41:163. 1928.

Andropogon scoparius var. *genuinus* Fern. & Grisc. Rhodora 37:145. 1935.

Schizachyrium scoparium var. *frequens* (Hubb.) Gould, Brittonia 19:73. 1967.

Loosely or densely tufted perennial; culms much branched, to 1.5 m tall; sheaths compressed, glabrous or villous; blades 3–7 mm broad, green or glaucous, glabrous or nearly so to puberulent throughout; racemes solitary, long-exsert, to 7 cm long, the rachis ciliate; sessile spikelets 4.5–8.0 mm long; awn of fertile lemma 7–14 mm long; pedicellate spikelet sterile, 3–5 mm long, short-awned, the pedicel densely ciliate; 2n = 40 (Hunter, 1934).

COMMON NAME: Little Bluestem.

HABITAT: Prairies; fields; open woodlands.

RANGE: New Brunswick to Alberta, south to Arizona, Texas, and Florida.

ILLINOIS DISTRIBUTION: Occasional to common throughout the state.

Little Bluestem is a highly variable species, with many of the extremes in variation having been named. In their extreme form, the variations seem rather distinct, but intergardation makes it impossible to segregate many specimens into satisfactory varieties. Although no attempt has been made in this work to assign each specimen studied to a particular variety (which is not possible), extreme specimens assignable to four different varieties have been encountered. Those with a partially bearded rachis and simple inflorescence with few branches have been called *Andropogon scoparius* var. *scoparius* (var. *villosissimus* is synonymous); those with a partially bearded rachis and much branched inflorescence have been called *A. scoparius* var. *polycladus;* those with a completely bearded rachis and glumes 4.5–6.0 mm long have been called *A. scoparius* var. *frequens;* while those with a completely bearded rachis and glumes over 6 mm long have been called *A. scoparius* var. *neomexicanus.*

Although some specimens may begin to flower as early as mid-June, most plants flower from late August to early October.

The segregation of Little Bluestem into the genus *Schizachyrium* is based primarily on inflorescence type. For further discussion, the reader is referred to Gould (1967).

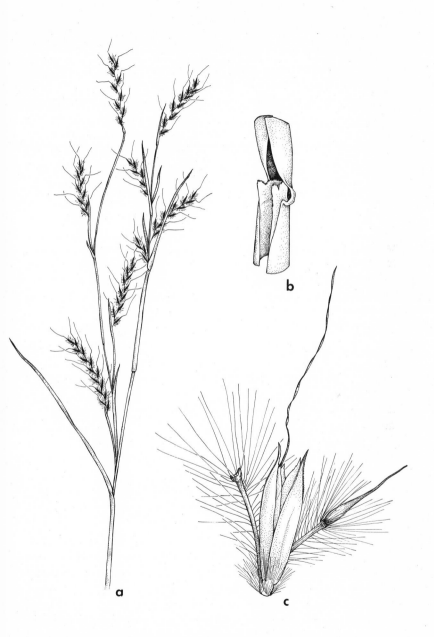

205. *Schizachyrium scoparium* (Little Bluestem). *a.* Inflorescences, X½. *b.* Sheath, with ligule, X5. *c.* Spikelet, X6.

Specimens collected in 1968 in Azotus Field, Pope County, are identifiable to *Andropogon praematurus* Fern. This binomial refers to specimens which resemble Little Bluestem except that they bear shorter racemes, a terminal spikelet subtended by two pedicels, and a single, staminate, pedicellate lateral spikelet. It has recently been discovered that *A. praematurus* merely represents a smut-infested form of some species of Bluestem. It is true that the Pope County specimens are heavily smut-infested.

58. *Tripsacum* L. – Gama Grass

Robust perennial; blades flat; inflorescence paniculate, monoecious, the pistillate part below; spikelets unisexual; staminate spikelets 2-flowered, borne in pairs on one side of the continuous rachis, one or both sessile; pistillate spikelets solitary, sunken on opposite sides of the jointed rachis below the staminate portion, composed of one perfect floret and a sterile lemma; staminate glumes membranous, the pistillate ones cartilaginous and broader; lemmas membranous and usually hyaline.

Only the following species occurs in Illinois.

1. **Tripsacum dactyloides** (L.) L. Syst. Nat. 2:1261. 1759.

Fig. 206.

Coix dactyloides L. Sp. Pl. 972. 1753.
Perennial from thick, creeping rhizomes; culms to 3 m tall; blades 15–30 mm broad; spikes to 25 cm long, pistillate for nearly one-third the length, the terminal 1–3 in number, the axillary solitary; spikelets 7–10 mm long; 2n = 36, 72 (Mangelsdorf & Reeves, 1939).

COMMON NAME: Gama Grass.
HABITAT: Low ground.
RANGE: Massachusetts to Nebraska, south to Texas and Florida; Mexico.
ILLINOIS DISTRIBUTION: Occasional in the southern two-thirds of the state; absent elsewhere.
This species somewhat resembles corn, to which it is closely related. It flowers from late May to early September.
Both *Tripsacum* and *Zea* have been assigned traditionally to tribe Tripsaceae on the basis of the unisexual spikelets. There appears to be justification, however, for combining tribe Tripsaceae with tribe Andropogoneae.

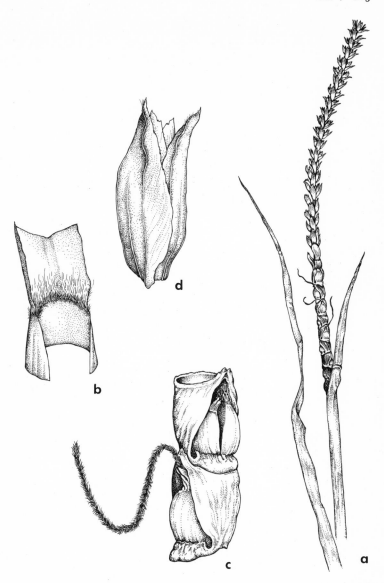

206. *Tripsacum dactyloides* (Gama Grass). *a.* Inflorescence, X½. *b.* Sheath, with ligule, X5. *c.* Pistillate spikelet, X4. *d.* Staminate spikelet, X4.

59. *Zea* L. – Corn

Robust annuals; blades flat, very broad; terminal panicle composed of staminate racemes (the tassels); axillary spikes pistil-

207. *Zea mays* (Corn). *a*. Staminate inflorescence, X½. *b*. Staminate spike-lets, X5.

late, enclosed in spathes (the ears); spikelets unisexual; staminate spikelets 2-flowered, borne in pairs on one side of the continuous rachis, one sessile or subsessile, the other pedicellate; glumes membranous; lemmas hyaline; 2n = 10, 40, 80 (Randolph, 1932), 20 (Darlington & Upcott, 1941).

Only the following species comprises the genus.

1. Zea mays L. Sp. Pl. 971. 1753. *Fig. 207.*

Robust annual to 5 m tall; staminate spikelets in long racemes, forming a spreading terminal panicle; pistillate inflorescence axillary, the spikelets in 8–25 rows borne on a woody axis (the cob); styles long, silky.

COMMON NAME: Corn.

Corn often escapes from cultivation but is rarely persistent. It probably has escaped, at one time or another, in every county in the state (not mapped).

SUBFAMILY Eragrostoideae

Annuals or perennials; inflorescence various; spikelets 1- to several-flowered, usually all fertile; lemmas mostly 3-nerved (1-nerved in *Calamovilfa* and *Sporobolus,* 7- to 11-nerved in *Distichlis*).

This subfamily includes several festucaceous genera of Hitchcock and many genera of Hitchcock's tribe Chlorideae. It also includes the introduced tribe Zoysieae.

Under the classification followed here, the subfamily Eragrostoideae is composed of the following tribes in Illinois: Eragrosteae, Chlorideae, Zoysieae, Aeluropodeae, and Aristideae.

Tribe *Eragrosteae*

Annuals or perennials; inflorescence mostly paniculate; spikelets 1- to several-flowered; disarticulation above the glumes (except in *Muhlenbergia*); lemmas 3-nerved (except *Sporobolus* and *Calamovilfa*).

Genera assigned to this tribe in the "new" system of classification are *Eragrostis, Tridens, Triplasis, Redfieldia, Calamovilfa, Muhlenbergia, Sporobolus,* and *Crypsis.* The characteristics used to group these genera into a single tribe are too difficult to observe and are not suitable for discussion in a flora.

60. *Eragrostis* BEAUV. – Love Grass

Annuals or perennials; blades usually flat or folded, rarely involute; inflorescence paniculate; spikelets compressed, 2- to many-flowered, disarticulating above the glumes; glumes 2, somewhat unequal, shorter than the spikelets; lemmas 3-nerved, keeled, without a tuft of cobwebby hairs at the base.

Of the thirteen species of *Eragrostis* known from Illinois, seven are native and six are adventive. The taxonomy of the genus is rather difficult, with measurements of closely related species tending to overlap. Examples of this may be found in the complex of *E. capillaris, E. pilosa,* and *E. frankii,* or between *E. pectinacea* and *E. diffusa.* Taxonomists are not in agreement concerning the status of several of the species (e.g., *E. diffusa, E. pilifera, E. neomexicana*). A revision of *Eragrostis* by Harvey (1948) has been consulted in the preparation of this treatment of the Illinois species.

Often the entire panicle is dislodged from the plant and blows about in the wind. Such species are known as tumbleweeds.

KEY TO THE SPECIES OF *Eragrostis* IN ILLINOIS

1. Plants forming mats, the culms rooting at the nodes.
    2. Sheaths more or less glabrous; inflorescence 2–8 cm long, the peduncle glabrous; lemmas 1.5–2.0 mm long, acute, glabrous; anthers 0.2–0.5 mm long; plants monoecious___1. *E. hypnoides*
    2. Sheaths pubescent; inflorescence 10–25 cm long, the peduncle villous; lemmas 2–4 mm long, acuminate, sparsely villous along the nerves; anthers 1.5–2.0 mm long; plants dioecious_____ _____2. *E. reptans*
1. Plants erect or ascending, not forming mats.
    3. All lemmas 2.5–3.5 mm long; perennials_____3. *E. trichodes*
    3. Lemmas (or most of them) 1.0–2.5 mm long; annuals (except nos. 6 and 7).
        4. Leaf-blades with small, rounded, wart-like projections (glands) along the margin; lemmas often glandular along the keel.
            5. Spikelets 2.2–3.0 mm broad; lemmas 2.0–2.5 mm long, glandular along the keel; anthers 0.5 mm long_____ _____4. *E. cilianensis*
            5. Spikelets 1.5–2.0 mm broad; lemmas 1.5–2.0 mm long, glandular or eglandular along the keel; anthers 0.2 mm long_____5. *E. poaeoides*

4. Leaf-blades and lemmas without wart-like projections (glands).
  6. Plants perennial from short rhizomes.
    7. Spikelets purplish, 3–8 mm long; branches of the panicle spreading or reflexed; blades flat or folded____ _____6. *E. spectabilis*
    7. Spikelets grayish-green, 8–10 mm long; branches of the panicle ascending; blades involute_____ _____7. *E. curvula*
  6. Plants annual, tufted, without rhizomes.
    8. Lemmas 2.0–2.5 mm long; second glume 1.8–2.0 mm long_____8. *E. neomexicana*
    8. Lemmas 1.0–1.8 mm long; second glume 0.8–1.6 mm long.
      9. Lemmas conspicuously 3-nerved; spikelets 4–8 mm long.
        10. Panicle 15–25 (–30) cm long_____ _____9. *E. pectinacea*
        10. Panicle 30–50 cm long_____10. *E. diffusa*
      9. Lemmas obscurely 3-nerved; spikelets 1.5–4.0 mm long.
        11. Panicle at least two-thirds as long as the entire plant; first glume 1.0–1.5 mm long_____ _____11. *E. capillaris*
        11. Panicle not more than one-half as long as the entire plant; first glume 0.5–1.2 mm long.
          12. Panicle broadest near base; second glume one-half to three-fourths as long as the lowest lemma_____12. *E. pilosa*
          12. Panicle broadest near middle; second glume as long as the lowest lemma_____ _____13. *E. frankii*

1. Eragrostis hypnoides (Lam.) BSP. Prel. Cat. N. Y. 69. 1888. *Fig. 208.*

Poa hypnoides Lam. Tabl. Encycl. 1:185. 1791.
Creeping, mat-forming, monoecious annual, rooting at the nodes; sheaths more or less glabrous; blades scabrous or puberulent above, 1–3 mm broad; inflorescence 2–8 cm long, the peduncles glabrous; spikelets 5–15 mm long, 1.0–2.5 mm broad, 10- to 35-

flowered; glumes subacute, the first usually a little less than 1 mm long, the second 1.0–1.8 mm long; lemmas 1.5–2.0 mm long, acute, glabrous; anthers 0.2–0.5 mm long.

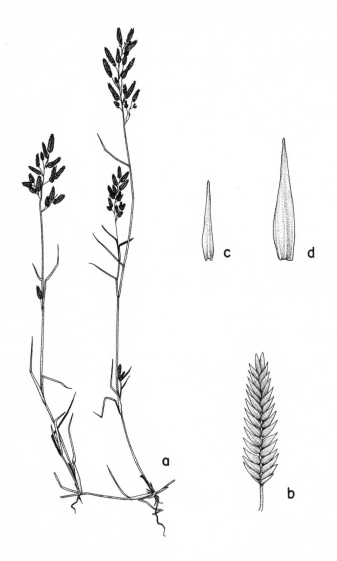

208. *Eragrostis hypnoides* (Pony Grass). *a*. Habit, X½. *b*. Spikelet, X4. *c*. First glume, X16. *d*. First lemma, X16.

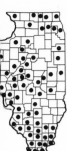

COMMON NAME: Pony Grass.

HABITAT: Wet ground, usually in sandy areas.

RANGE: Quebec to Washington, south to California, Texas, and Florida; West Indies; Mexico; South America.

ILLINOIS DISTRIBUTION: Not uncommon; throughout the state. This species may form dense mats occupying several square feet. It flowers from mid-July into October. It is distinguished easily from *E. reptans,* the other creeping *Eragrostis,* by its more glabrous nature, its smaller lemmas, and its monoecious condition.

2. **Eragrostis reptans** (Michx.) Nees, Agrost. Bras. 514. 1829.

Fig. 209.

Poa reptans Michx. Fl. Bor. Am. 1:69. 1803.

Poa dioica Michx. ex Poir. in Lam. Encycl. 5:87. 1804.

Poa weigeltiana Reichenb. ex Trin. Mem. Acad. St. Petersb. VI. Math. Phys. Nat. 1:410. 1830.

Eragrostis weigeltiana (Reichenb.) Bush, Trans. Acad. Sci. St. Louis 13:180. 1903.

Neeragrostis reptans (Michx.) Nicora, Rev. Argent. Agron. 29:5. 1962.

Creeping, mat-forming, dioecious annual, rooting at the nodes; sheaths pubescent; blades pubescent, 1–3 mm broad; inflorescence 10–25 cm long, the peduncles villous; spikelets 5–18 mm long, 2–4 mm broad, 10- to 75-flowered; glumes acute, the first 1.0–1.5 mm long, the second 1.5–2.2 mm long; lemmas 2–4 mm long, acuminate, sparsely villous along the nerves; anthers 1.5–2.0 mm long.

COMMON NAME: Pony Grass.

HABITAT: Sandy soil.

RANGE: Kentucky to South Dakota, south to Texas and Tennessee; Mexico.

ILLINOIS DISTRIBUTION: Occasional; scattered throughout the state.

This species, unique among the Illinois species of *Eragrostis* in being dioecious, flowers from mid-July to mid-October. It is far less common than *E. hypnoides,* the other mat-forming species in Illinois.

The type was collected by Michaux from what is now Randolph County.

Nicora (1962) proposes that this species be segregated into its own genus, called *Neeragrostis,* on the basis of its dioecious condition and technical characters of the ovary and the leaf epidermis. She postulates a relationship with the genus *Distichlis.* It is my opinion, however, that these differences are not sufficient enough to merit generic segregation from *Eragrostis.*

3. Eragrostis trichodes (Nutt.) Wood, Class-book 796. 1861.

Poa trichodes Nutt. Trans. Amer. Phil. Soc. 5:146. 1837.
Cespitose perennial with slender culms to 1.2 m tall; sheaths glabrous, except for the pilose summit; blades rigid, involute at the tip, 2–7 mm broad; inflorescence 20–60 cm long, spreading to ascending, purplish or yellowish; spikelets 4–12 mm long, 3- to 15-flowered; glumes acute or acuminate, the first 1.5–3.2 mm long, the second 2.0–3.5 mm long; lemmas subulate, 2.5–3.5 mm long; anthers 1.0–1.2 mm long.

Two well-marked varieties occur in Illinois.

1. Inflorescence purplish; spikelets 3- to 6-flowered, 4–7 mm long; glumes and lemmas to 3 mm long___3a. *E. trichodes* var. *trichodes*
1. Inflorescence yellowish; spikelets 8- to 15-flowered, 8–12 mm long; glumes and lemmas 3.0–3.5 mm long_____
_____3b. *E. trichodes* var. *pilifera*

3a. Eragrostis trichodes (Nutt.) Wood var. **trichodes** *Fig. 210.*

Eragrostis tenuis Gray, Man. ed. 6, 661. 1890, non Steud. (1854).
Leaves 2–6 mm broad; inflorescence 20–50 cm long, purplish; spikelets 4–7 mm long, 3- to 6-flowered; first glume 1.5–3.0 mm long; anthers 1 mm long.

COMMON NAME: Thread Love Grass.

HABITAT: Open, sandy areas.

RANGE: Ohio to Nebraska, south to Texas and Louisiana.

ILLINOIS DISTRIBUTION: Occasional; restricted to the western half of the state.

This taxon is a common inhabitant of the Illinois sand prairies so typical along the Illinois River. It flowers from July to mid-October.

It is smaller in all respects from var. *pilifera.* Some authors have considered *E. pilifera* to be a species dis-

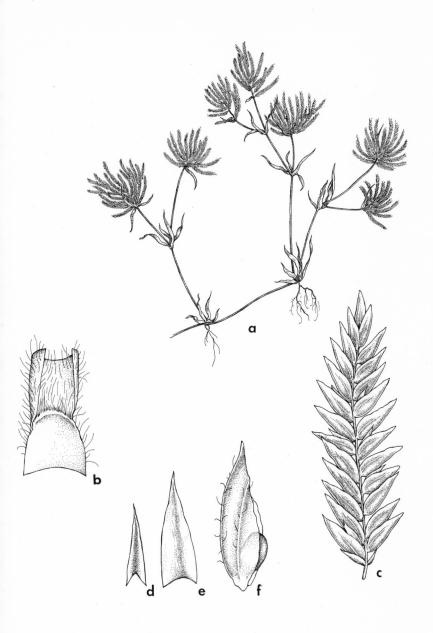

209. *Eragrostis reptans* (Pony Grass). *a.* Habit, X½. *b.* Sheath, with ligule, X7½. *c.* Pistillate spikelet, X5. *d.* First glume, X20. *e.* Second glume, X20. *f.* Lemma, X20.

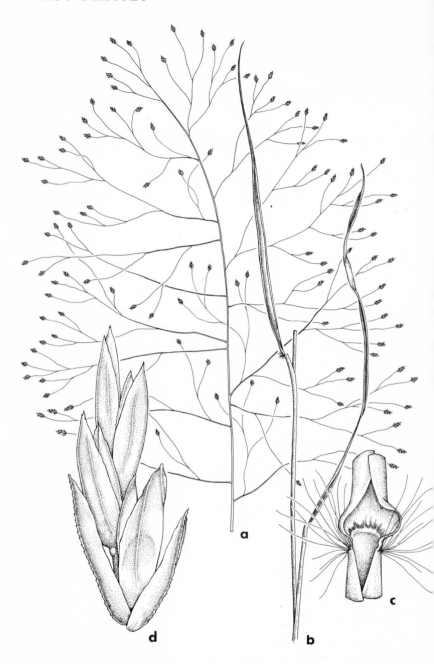

210. *Eragrostis trichodes* var. *trichodes* (Thread Love Grass). *a.* Inflorescence, X½. *b.* Leaves, X½. *c.* Sheath, with ligule, X7½. *d.* Spikelet, X15.

tinct from *E. trichodes*. In addition to size difference, the inflorescence is purple in var. *trichodes* and yellow in var. *pilifera*.

3b. Eragrostis trichodes (Nutt.) Wood var. **pilifera** (Scheele)

Fern. Rhodora 40:331. 1938. *Fig. 211.*

Eragrostis pilifera Scheele, Linnaea 22:344. 1849.
Eragrostis grandiflora Smith & Bush, Rept. Mo. Bot. Gard. 6:117. 1895.

Leaves 2–7 mm broad; inflorescence 25–60 cm long, yellowish; spikelets 8–12 mm long, 8- to 15-flowered; first glume 2.0–3.2 mm long, the second glume 3.0–3.5 mm long; lemmas 3.0–3.5 mm long; anthers 1.2 mm long.

COMMON NAME: Thread Love Grass.
HABITAT: Sand prairies.
RANGE: Illinois to Nebraska, south to Texas and Louisiana.
ILLINOIS DISTRIBUTION: Not common; confined to sand prairies along the Illinois River.
This taxon is larger in most respects than var. *trichodes*. Chase (1951) maintains *E. pilifera* as a distinct species.

4. Eragrostis cilianensis (All.) Mosher, Bull. Ill. Agr. Exp.

Sta. 205:381. 1918. *Fig. 212.*

Briza eragrostis L. Sp. Pl. 70. 1753.
Poa cilianensis All. Fl. Pedem. 2:246. 1785.
Poa megastachya Koel. Descr. Gram. 181. 1802.
Eragrostis major Host, Icon. Gram. Austr. 4:14. 1809.
Eragrostis megastachya (Koel.) Link, Hort. Berol. 1:187. 1827.
Eragrostis poaeoides var. *megastachya* (Koel.) Gray, Man., ed. 2, 563. 1856.

Densely cespitose annual, decumbent at the base, with culms to 75 cm tall; sheaths glabrous below, pilose above; blades glandular along the margins, 2–7 mm broad; inflorescence 5–17 mm long, 2.2–3.0 mm broad, 10- to 40-flowered, dark greenish; glumes acute or subacute, the first 1.2–2.0 mm long, the second 1.5–2.0 mm long; lemmas narrowly ovate, 2.0–2.5 mm long, scabrous, glandular along the keel; anthers 0.5 mm long; 2n = 20 (Avdulov, 1928).

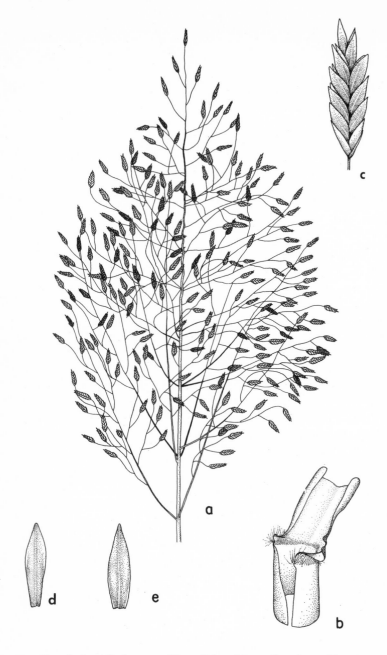

211. *Eragrostis trichodes* var. *pilifera* (Thread Love Grass). *a.* Upper part of plant, X½. *b.* Sheath, with ligule, X6. *c.* Spikelet, X4. *d.* Second glume, X12. *e.* Fourth lemma, X12.

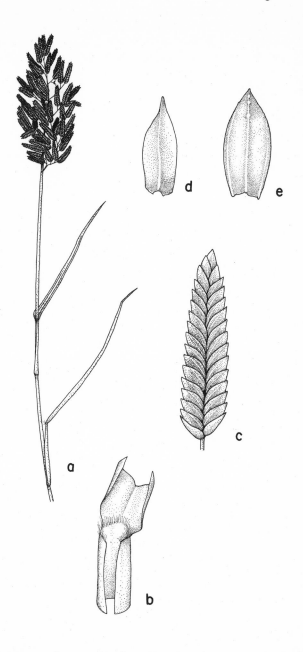

212. Eragrostis cilianensis (Stinking Love Grass). *a.* Upper part of plant, X½. *b.* Sheath, with ligule, X6. *c.* Spikelet, X4. *d.* Second glume, X12. *e.* Fourth lemma, X12.

COMMON NAME: Stinking Love Grass.
HABITAT: Fields, waste ground.
RANGE: Native of Europe and Asia; introduced throughout the United States.
ILLINOIS DISTRIBUTION: Common; in every county.
Considerable variation exists in the size and number of flowers per spikelet. Largest spikelets measure about 17 mm long and have up to 40 florets; the smallest spikelets seen from Illinois were 5 mm long and 10-flowered.

Some confusion exists as to the correct name of this plant. It was described by Linnaeus as *Briza eragrostis*, but the epithet *eragrostis* cannot be transferred to the genus *Eragrostis*. The next earliest name is *Poa cilianensis*, although Fernald (1950) doubts the identity of this binomial and suggests that *E. megastachya* be used for our species. Chase (1951) maintains *E. cilianensis*, however.

There is sometimes difficulty in distinguishing *E. cilianensis* from *E. poaeoides*. Most reliable characters in separating these two species are the length and width of the spikelets and the size of the anthers.

The common name of Stinking Love Grass alludes to the disagreeable odor of the fresh plants. The flowers are produced from late May to early October.

5. Eragrostis poaeoides Beauv. ex Roem. & Schultes, Syst. Veg. 2:574. 1817. *Fig. 213.*

Poa eragrostis L. Sp. Pl. 68. 1753.
Eragrostis minor Host. Icon. Gram. Austr. 4:15. 1809. (Illegitimate since *Eragrostis* was not a validly published genus until 1812.)
Eragrostis eragrostis (L.) Beauv. Ess. Agrost. 71, 174. 1812.
Cespitose annual, more slender than *E. cilianensis*, with culms to 50 cm tall; sheaths glabrous below, pilose above; blades glandular along the margins, 2–5 mm broad; inflorescence 3–10 cm long, spreading, rarely suberect; spikelets 5–10 mm long, 1.5–2.0 mm broad, 8- to 20-flowered; glumes acute or subacute, the first 1.2–1.5 mm long, the second 1.4–1.7 mm long; lemmas narrowly ovate, 1.5–2.0 mm long, minutely glandular or eglandular along the keel; anthers 0.2 mm long.

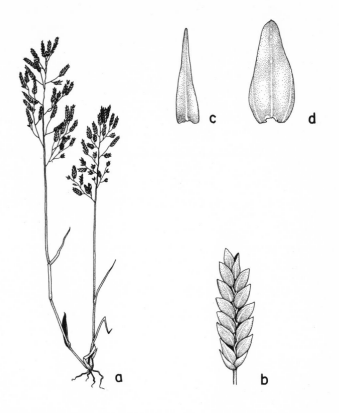

213. *Eragrostis poaeoides* (Love Grass). *a.* Habit, X½. *b.* Spikelet, X4. *c.* Second glume, X16. *d.* Lemma, X16.

COMMON NAME: Love Grass.

HABITAT: Fields, waste ground.

RANGE: Native of Europe; introduced throughout the United States.

ILLINOIS DISTRIBUTION: Common; probably in every county.

This species is similar to *E. cilianensis* in that both have glandular margins along the leaf blades. Unlike *E. cilianensis*, however, *E. poaeoides* does not always have glands on the keel of the lemma. *Eragrostis poaeoides* is generally smaller than *E. cilianensis* in all respects.

This species flowers from early June to early October.

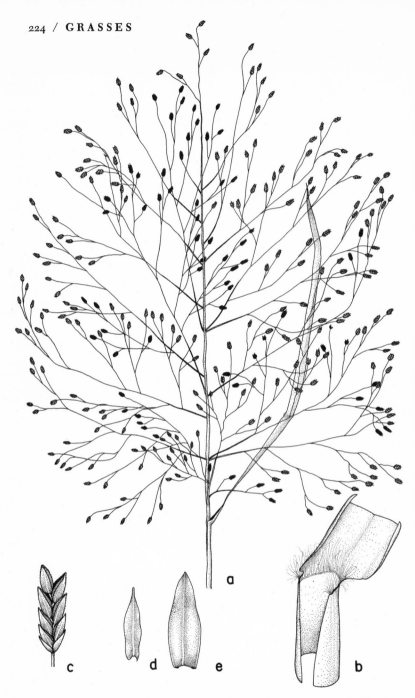

214. *Eragrostis spectabilis* (Tumble-grass). *a.* Inflorescence, X½. *b.* Sheath, with ligule, X6. *c.* Spikelet, X5. *d.* Second glume, X16. *e.* Lemma, X16.

6. Eragrostis spectabilis (Pursh) Steud. Nom. Bot., ed. 2, 1:564. 1840. *Fig. 214.*

Poa spectabilis Pursh, Fl. Amer. Sept. 1:81. 1814.
Eragrostis geyeri Steud. Syn. Pl. Glum. 1:272. 1854.
Eragrostis pectinacea var. *spectabilis* (Pursh) Gray, Man. ed. 2, 565. 1856.
Eragrostis spectabilis var. *sparsihirsuta* Farw. Am. Midl. Nat. 10:306. 1927.

Cespitose perennial from short rhizomes, with culms to 75 cm tall; sheaths glabrous or pilose below, densely hairy at the throat; blades firm, flat or folded, glabrous or rarely pilose, 3–8 mm broad; inflorescence 10–45 cm long, spreading or reflexed, purple; spikelets 3–8 mm long, 1.5–2.0 mm broad, 3- to 12-flowered; glumes acute, the first 1.0–1.8 mm long, the second 1.2–2.0 mm long; lemmas 1.5–2.5 mm long, ovate, obtuse to sub-acute, scabrous on the keel.

COMMON NAME: Tumble-grass.
HABITAT: Sandy soil.
RANGE: Massachusetts to Minnesota, south to Texas and Florida.
ILLINOIS DISTRIBUTION: Common; in every county.
This handsome grass flowers from late June well into October. The entire panicle often breaks away from the plant and is blown about by the wind. The common name Tumble-grass is very appropriate.

There is variation with the pubescence of the sheaths and blades, but the majority of specimens have generally glabrous blades. Those with pilose sheaths have been called var. *sparsihirsuta*. Number of flowers per spikelet is variable, but the predominant number in Illinois seems to be between six and nine. There is rather wide variation in length of glumes and lemmas.

For several years, this species erroneously was called *E. pectinacea*, but *E. pectinacea* is an entirely different annual species.

Eragrostis geyeri, which is synonymous with *E. spectabilis*, was described from material collected by Geyer in Cass County.

7. **Eragrostis curvula** (Schrad.) Nees, Fl. Afr. Austr. 397. 1841. *Fig. 215.*

Poa curvula Schrad. Gott. Anz. Ges. Wiss. 3:2073. 1821.
Densely tufted perennial to 1.2 m tall; sheaths keeled, the lower densely hairy, the upper glabrous or sparsely hispid; blades involute, scabrous; inflorescence 20–30 cm long, ascending; spikelets 8–10 mm long, 7- to 11-flowered, grayish-green; glumes subacute, 1.5–2.0 mm long; lemmas 1.8–2.5 mm long, obtuse to subacute.

COMMON NAME: Weeping Love Grass.
HABITAT: Waste ground.
RANGE: Native of Africa; gradually becoming established in the United States, particularly in the south and southwest.
ILLINOIS DISTRIBUTION: Rare; known only from Morgan County (east of Meredosia, September 21, 1960, *R. T. Rexroat 7214, 7214A*).
The larger, grayish-green spikelets and the involute blades readily distinguish *E. curvula* from *E. spectabilis*.

8. **Eragrostis neomexicana** Vasey, Contrib. U. S. Nat. Herb. 2:542. 1894. *Fig. 216.*

Stout annual with culms to 1 m tall; sheaths glabrous below, pilose at the throat; blades 3–10 mm broad; inflorescence 15–40 cm long, spreading or ascending; spikelets 5–8 mm long, 1.8–2.9 mm broad, 8- to 12-flowered, dark grayish-green; glumes acute, the first 1.4–1.7 mm long, the second 1.8–2.0 mm long; lemmas acute, 2.0–2.5 mm long, scabrous on the keel.

COMMON NAME: Love Grass.
HABITAT: Waste ground.
RANGE: Native in the southwestern United States; casually adventive eastward.
ILLINOIS DISTRIBUTION: Rare; known only from St. Clair County (along railroad, E. St. Louis, July 17, 1964, *Mohlenbrock 14222*).
The taxonomy of this species is not agreed upon by all botanists. Its distinctness from *E. mexicana*, which is not known from Illinois, is questionable. It differs from

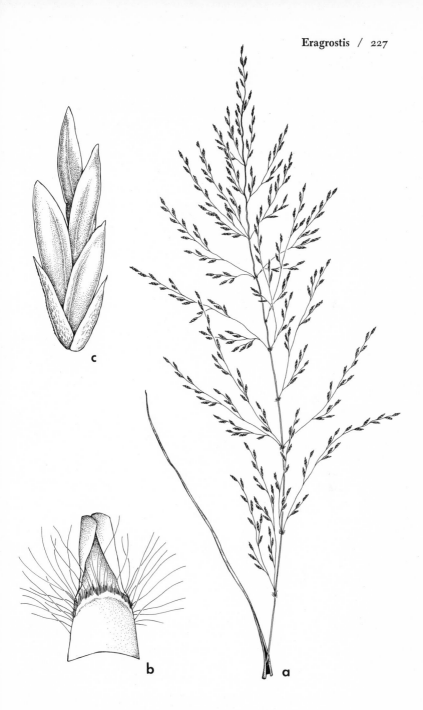

215. *Eragrostis curvula* (Weeping Love Grass). *a.* Inflorescence, X½. *b.* Sheath, with ligule, X7½. *c.* Spikelet, X7½.

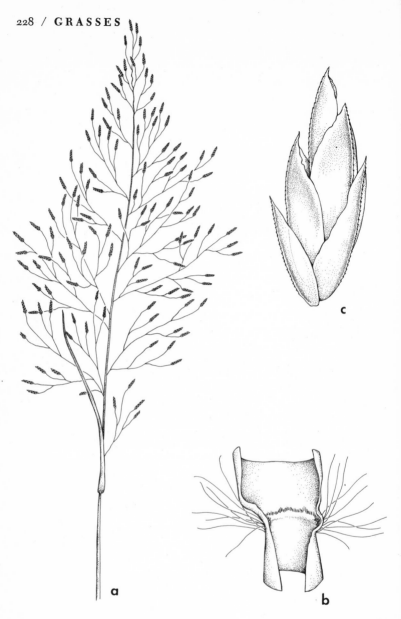

216. *Eragrostis neomexicana* (Love Grass). *a.* Inflorescence, X½. *b.* Sheath, with ligule, X7½. *c.* Spikelet, X10.

similar Illinois species such as *E. pectinacea* and *E. diffusa* by its longer lemmas and second glumes.

This is one of the most robust species of *Eragrostis* in Illinois.

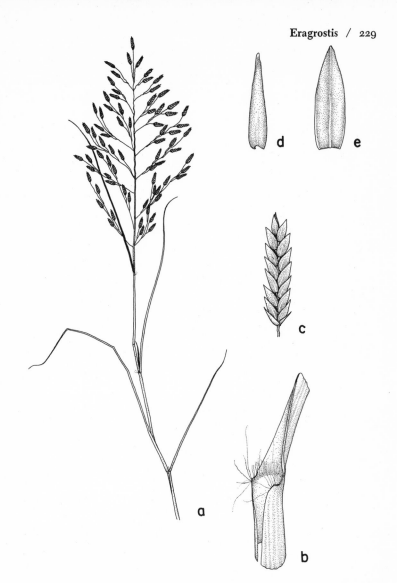

217. *Eragrostis pectinacea* (Love Grass). *a.* Upper part of plant, X½. *b.* Sheath, X6. *c.* Spikelet, X4. *d.* Second glume, X16. *e.* Lemma, X16.

9. **Eragrostis pectinacea** (Michx.) Nees, Fl. Afr. Austr. 406. 1841. *Fig. 217.*

Poa pectinacea Michx. Fl. Bor. Amer. 1:69. 1803.

Densely cespitose annual, often decumbent at the base, with slender culms to 75 cm tall; sheaths glabrous below, hairy at the summit; blades 1–3 mm broad; inflorescence 15–25 cm long, spreading; spikelets 4–8 mm long, 1.0–1.8 mm broad, 5- to 15-flowered, lead-colored; glumes acuminate, scabrous on the keel, the first 0.8–1.2 mm long, the second 1.0–1.5 mm long; lemmas acute, the lowest 1.5–1.8 mm long, prominently 3-nerved.

COMMON NAME: Love Grass.
HABITAT: Waste ground, fields.
RANGE: Quebec to British Columbia, south to California, Texas, and Florida; Mexico.
ILLINOIS DISTRIBUTION: Common; in every county.
The Michaux type is from Illinois.
This species flowers from July to early October. Its smaller lemmas distinguish it from *E. neomexicana,* while its conspicuously nerved lemmas separate it from *E. capillaris, E. pilosa,* and *E. frankii.*
Variation may be seen in the width of the blades and the number of flowers per spikelet.

10. **Eragrostis diffusa** Buckl. in Proc. Acad. Nat. Sci. Phila. 1862:97. 1862. *Fig. 218.*

Cespitose annual with culms to 50 cm tall; sheaths glabrous below, hairy at the throat; blades 1–3 mm broad; inflorescence 30–50 cm long, with the primary branches bearing numerous secondary ones; spikelets 4–10 mm long, 5- to 15-flowered; glumes acuminate, the first 0.8–1.2 mm long, the second 1.0–1.6 mm long; lemmas acute, 1.5–1.6 mm long, prominently 3-nerved.

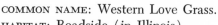

COMMON NAME: Western Love Grass.
HABITAT: Roadside (in Illinois).
RANGE: Native in the western United States; occasionally adventive eastward.
ILLINOIS DISTRIBUTION: Rare; known only from Menard County (*Lansing & Sherff* in 1916).
This species is similar to *E. pectinacea* and combined with it by many authors. *Eragrostis diffusa* has a more dense inflorescence because of the presence of secondary branches along the primary panicle branches. Although measurements of glumes, lemmas, *et cetera* are nearly identical between *E. diffusa* and *E. pectinacea,* the two species have quite a different appearance.

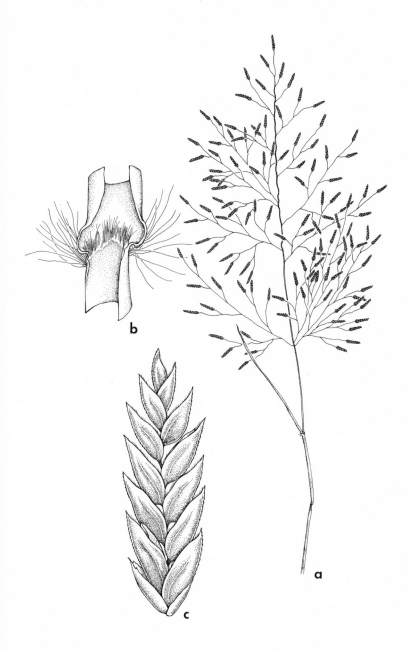

218. *Eragrostis diffusa* (Western Love Grass). *a.* Inflorescence, X½. *b.* Sheath, with ligule, X7½. *c.* Spikelet, X10.

219. *Eragrostis capillaris* (Lace Grass). *a.* Inflorescence, X½. *b.* Sheath, X6. *c.* Spikelet, X8. *d.* Glume, X16. *e.* Lemma, X16.

11. **Eragrostis capillaris** (L.) Nees, Agrost. Bras. 505. 1829.
Fig. 219.

Poa capillaris L. Sp. Pl. 68. 1753.
Cespitose annual with capillary culms to 70 cm tall; sheaths pilose
or glabrous below, long-hairy at the throat; blades pilose above
near the base, 2–4 mm broad; inflorescence 10–30 cm long,
spreading; spikelets 1.5–3.0 mm long, 1–2 mm broad, 2- to 5-
flowered, lead-colored; glumes acute, 1.2–1.8 mm long, scabrous
toward the tip, faintly 3-nerved.

COMMON NAME: Lace Grass.
HABITAT: Dry, rocky soil, particularly in woodlands.
RANGE: Maine to Wisconsin, south to Texas and Georgia.
ILLINOIS DISTRIBUTION: Occasional; scattered throughout
the state.
The capillary culms and panicle branches account for
the common name of Lace Grass. The panicle is at least
two-thirds as long as the entire plant. The flowering
period is July to September.

12. **Eragrostis pilosa** (L.) Beauv. Ess. Agrost. 71, 162, 175.
1812. *Fig. 220.*

Poa pilosa L. Sp. Pl. 68. 1753.
Annual with slender culms to 60 cm tall; sheaths glabrous or
pilose, long-hairy at the summit; blades 1–3 mm broad; inflorescence 5–20 cm long, spreading or ascending; spikelets 3–5
mm long, 1.0–1.2 mm broad, 3- to 9-flowered; glumes acute, the
first 0.5–1.0 mm long, the second 0.8–1.3 mm long; lemmas acute,
1.3–1.6 mm long, faintly nerved.

COMMON NAME: Love Grass.
HABITAT: Waste ground.
RANGE: Native of Europe; introduced in the United
States from Massachusetts to Colorado, south to Texas
and Florida.
ILLINOIS DISTRIBUTION: Not common; apparently re-
stricted to a few southern counties.
The specimens from Illinois fall about equally between
those with glabrous sheaths and those with pilose
sheaths.

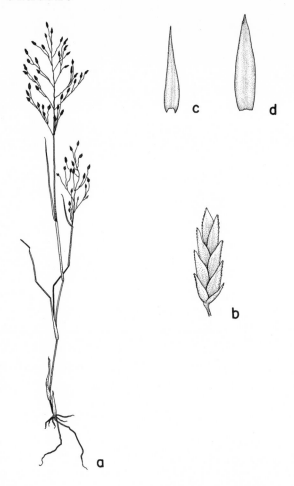

220. *Eragrostis pilosa* (Love Grass). *a.* Habit, X½. *b.* Spikelet, X8. *c.* Second glume, X16. *d.* Lemma, X16.

This species resembles *E. frankii,* but generally has smaller glumes and lemmas.

13. Eragrostis frankii C. A. Meyer ex Steud. Syn. Pl. Glum. 1:273. 1854.

Annual, more or less decumbent at the base, with slender culms to 45 cm tall; sheaths glabrous below, pilose at the throat; blades scabrous above, 1–4 mm broad; inflorescence 5–15 cm long, spreading or ascending; spikelets 2–4 mm long, 2- to 7-flowered; glumes acute, the first 0.8–1.2 mm long, the second 1.0–1.5 mm long, faintly nerved.

Two varieties may be recognized in Illinois.

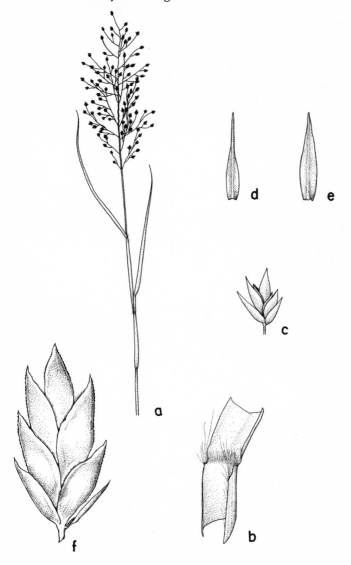

221. *Eragrostis frankii.*—var. *frankii.* *a.* Upper part of plant, X½. *b.* Sheath, X6. *c.* Spikelet, X8. *d.* Glume, X16. *e.* Lemma, X16. var. *brevipes.* *f.* Spikelet, X12½.

1. Spikelets 2- to 5-flowered, 2–3 mm long_____
_____13a. *E. frankii* var. *frankii*
1. Spikelets 6- to 7-flowered, 3–4 mm long_____
_____13b. *E. frankii* var. *brevipes*

13a. Eragrostis frankii C. A. Meyer ver frankii *Fig. 221a–e.*

Spikelets 2- to 5-flowered, 2–3 mm long.

HABITAT: Sandy soil.
RANGE: Quebec to Minnesota, south to Kansas, Louisiana, and Florida.
ILLINOIS DISTRIBUTION: Occasional; scattered throughout the state.

13b. Eragrostis frankii C. A. Meyer var. brevipes Fassett, Rhodora 34:95. 1932. *Fig. 221f.*

Spikelets 6- to 7-flowered, 3–4 mm long.

HABITAT: Sandy soil.
RANGE: Wisconsin and Illinois.
ILLINOIS DISTRIBUTION: Known only from Henderson County (Mississippi bottoms near Oquawka, *H. N. Patterson s.n.*).

61. *Tridens* ROEM. & SCHULT.

Tufted perennials; blades flat; inflorescence paniculate; spikelets 3- to 10-flowered, disarticulating above the glumes; glumes 2, more or less equal, shorter than the spikelets; lemmas rounded on the back, 3-nerved, pubescent at least on the nerves, mucronate between the apical teeth.

Although the two species of *Tridens* which occur in Illinois do not resemble each other greatly, they are placed in the same genus on the basis of the lemmas which have three pubescent or puberulent nerves. Many botanists have used the generic name *Triodia* for these species, but this name should be preserved primarily for Australian species.

KEY TO THE SPECIES OF Tridens IN ILLINOIS

1. Panicle loose, open; glumes oblong to ovate; lemmas 3.5–4.0 mm long_____1. *T. flavus*
1. Panicle contracted, spike-like; glumes linear-lanceolate; lemmas 2.5–3.0 mm long_____2. *T. strictus*

1. **Tridens flavus** (L.) Hitchcock, Rhodora 8:210. 1906. *Fig.* 222.

Poa flava L. Sp. Pl. 68. 1753.
Poa sesleroides Michx. Fl. Bor. Amer. 1:68. 1803.
Tricuspis sesleroides (Michx.) Torr. Fl. N. & Mid. U. S. 118. 1823.
Triodia sesleroides (Michx.) Benth. ex Vasey, Spec. Rept. U.S.D.A. 63:35. 1883
Triodia flava (L.) Smyth, Trans. Kans. Acad. 25:95. 1913.
Triodia flava f. *cuprea* Fosberg, Castanea. 11:67. 1946.

Tufted perennial with culms to 1.5 m tall; sheaths glabrous below, pubescent near summit; blades flat or becoming involute at apex, to 10 mm broad, glabrous; inflorescence 20–40 cm long, spreading or drooping; spikelets 5–10 mm long, 4- to 9-flowered, barely or not at all compressed, purple or yellow; glumes subequal, oblong to ovate, obtuse and mucronate, 2.5–3.5 mm long, 1-nerved; lemmas oblong, retuse or obtuse at apex, 3.5–4.0 mm long, villous in the lower half.

COMMON NAME: Purple-top.
HABITAT: Fields and edges of woodlands.
RANGE: New Hampshire to Minnesota, south to Texas and Florida; Central America.
ILLINOIS DISTRIBUTION: Common, except in the northern one-fifth of the state where it is apparently absent except in Cook County.
All but two Illinois specimens seen have purple spikelets, rather than yellow. The purple form has been segregated as f. *cuprea*. Specimens with yellow spikelets have been collected in Johnson and Sangamon counties.

Purple-top flowers from mid-June to the end of September.
The loose, inflorescence readily distinguishes this species from *T. strictus*.

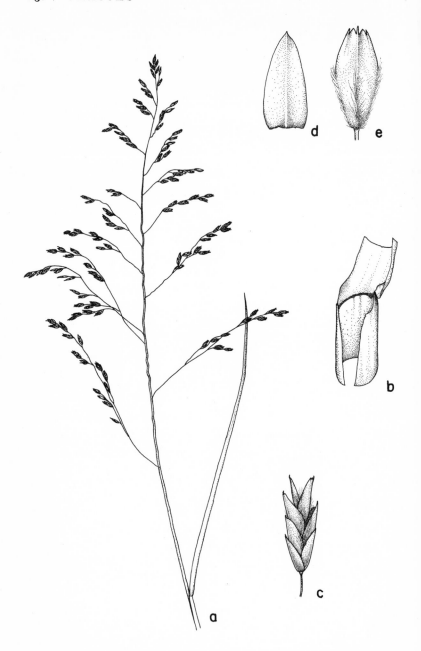

222. *Tridens flavus* (Purple-top). *a.* Inflorescence, X½. *b.* Sheath, X2. *c.* Spikelet, X4. *d.* Second glume, X8. *e.* Second lemma, X8.

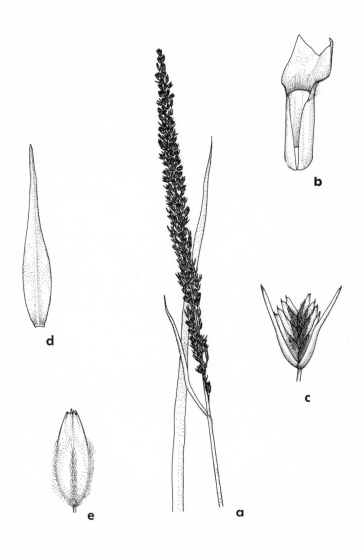

223. *Tridens strictus.* *a.* Inflorescence, X½. *b.* Sheath, X2. *c.* Spikelet, X5. *d.* First glume, X8. *e.* Third lemma, X8.

2. Tridens strictus (Nutt.) Nash in Small, Fl. S.E.U.S. 143. 1903. *Fig. 223.*

Windsoria stricta Nutt. Trans. Amer. Phil. Soc. 5:147. 1837.
Tricuspis stricta (Nutt.) Wood, Class-book 792. 1861.
Triodia stricta (Nutt.) Benth. ex Vasey, Spec. Rept. U.S.D.A. 63:35. 1883.

Densely tufted perennial to 1.5 m tall; sheaths glabrous; blades to 10 mm broad, glabrous; inflorescence contracted, spike-like, 10–30 cm long, to 2 cm thick; spikelets 4–6 mm long, 4- to 10-flowered, compressed, purplish; glumes subequal, linear-lanceolate, long-acuminate, 3–5 mm long, 1-nerved; lemmas oblong to ovate, obtuse or emarginate, 2.5–3.0 mm long, pubescent on the nerves below; 2n = 32 (Brown, 1950).

HABITAT: Waste ground and fields.
RANGE: Tennessee to Kansas, south to Texas and Alabama.
ILLINOIS DISTRIBUTION: Not common; scattered.
Except for the purple spikelets and characters of the lemmas, this species in no way resembles *T. flavus*. The dense, contracted, spike-like panicles particularly set this species apart from *T. flavus*.
This grass flowers from mid-July to October.

62. Triplasis BEAUV. – Sand Grass

Annuals (in Illinois) or perennials; sheaths loose; blades flat or involute; inflorescence paniculate, often partly enclosed by the enlarged sheaths; spikelets 2- to 5-flowered, disarticulating above the glumes; glumes 2, more or less equal, 1-nerved; lemmas rounded on the back, 3-nerved, bifid at apex, with a short awn between the teeth.

Only the following species occurs in Illinois.

1. Triplasis purpurea (Walt.) Chapm. Fl. South. U. S. 560. 1860. *Fig. 224.*

Aira purpurea Walt. Fl. Carol. 78. 1788.
Tricuspis purpurea (Walt.) Gray, Man. 589. 1848.

Tufted annual with wiry culms to nearly 1 m tall; sheaths scabrous; blades scabrous, 1–3 mm broad; inflorescence to 8 cm long, the stiff branches more or less ascending; spikelets 5–8 mm long, 2- to 5-flowered, purplish; glumes linear-lanceolate, acute, 2–4 mm long; lemmas 3–4 mm long, 3-nerved, villous, the awn 0.5–1.5 mm long, pubescent.

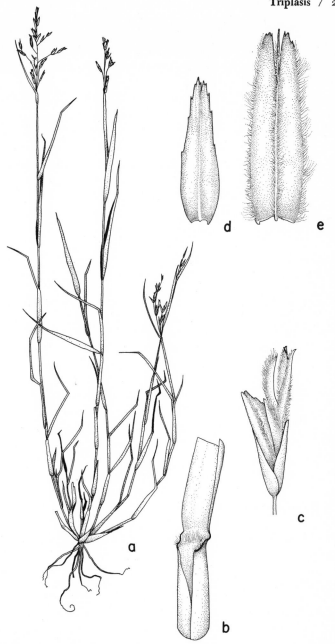

224. Triplasis purpurea (Sand Grass). *a.* Habit, X½. *b.* Sheath, X4. *c.* Spikelet, X6. *d.* Second glume, X16. *e.* First lemma, X16.

COMMON NAME: Sand Grass.

HABITAT: Sandy soil.

RANGE: Ontario to Minnesota, south to Colorado, Texas, and Florida.

ILLINOIS DISTRIBUTION: Occasional; absent from the southern one-fourth of the state.

This is another of the Illinois grasses, along with *Eragrostis trichodes*, *Ammophila breviligulata*, and *Calamovilfa longifolia* var. *magna*, restricted to sandy areas. It flowers from August to early October.

The villous, 3-nerved, awned lemmas, which measure 3–4 mm in length, are the chief distinguishing features of this genus.

63. *Redfieldia* VASEY – Blowout Grass

Perennials from slender, creeping rhizomes; blades elongate, involute; inflorescence large, paniculate; spikelets few-flowered, disarticulating above the glumes; glumes 2, more or less unequal, shorter than the spikelets; lemmas chartaceous, 3-nerved.

Only the following species comprises the genus.

1. Redfieldia flexuosa (Thurb.) Vasey, Bull. Torrey Club 14:133. 1887. *Fig. 225.*

Graphephorum flexuosum Thurb. in Gray, Proc. Acad. Nat. Sci. Phila. 1863:78. 1863.

Culms wiry, to 1 m tall; blades involute, elongate, glabrous; inflorescence 20–50 cm long, spreading to ascending, the branches capillary; spikelets 5–7 mm long, 3- to 4-flowered; glumes 1-nerved, acuminate, the first 2–3 mm long, the second 2.5–3.5 mm long; lemmas 3-nerved, acute and sometimes mucronate, 4–5 mm long, glabrous except at the base.

COMMON NAME: Blowout Grass.

HABITAT: Sand hills (in its native habitat).

RANGE: South Dakota to Utah, south to Arizona and Oklahoma; adventive in Illinois.

ILLINOIS DISTRIBUTION: Rare; adventive in Hancock County.

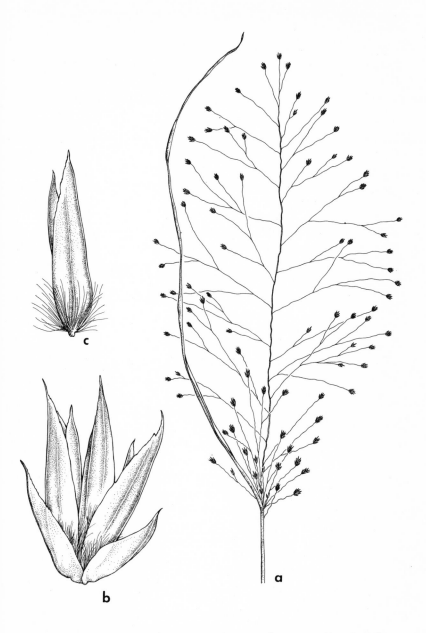

225. *Redfieldia flexuosa* (Blowout Grass). *a.* Inflorescence, X½. *b.* Spikelet, X10. *c.* Lemma, X9.

64. Calamovilfa (Gray) HACK. – Sand Reed

Perennials from creeping rhizomes; blades involute; inflorescence paniculate, spreading; spikelets 1-flavored, disarticulating above the glumes; glumes unequal, papery; lemmas rounded on the back, obscurely nerved, awnless, with a bearded callus.

Only the following species occurs in Illinois.

51. Calamovilfa longifolia (Hook.) Scribn. in Hack. var. magna Scribn. & Merrill, U.S.D.A. Div. Agrost. Circ. 35:3. 1901. *Fig. 226.*

Perennial from stout, scaly rhizomes, with culms to 1.5 m tall; sheaths appressed-pubescent, at least at the apex; blades flat at base, 3–8 mm broad, becoming involute above and capillary at the tip; panicle spreading-ascending, to 77.5 cm long; spikelets 6–7 mm long; glumes papery, glabrous, 1-nerved, acute to acuminate, the first lanceolate, 3.5–6.0 mm long, the second ovate, 4.5–7.5 mm long; lemma papery, ovate, 1-nerved, 4–7 mm long, glabrous except for the bearded callus.

COMMON NAME: Sand Reed.

HABITAT: Sandy areas.

RANGE: Shores of Lake Michigan and Lake Huron; northwestern Indiana; Illinois; southwestern Wisconsin; northeastern Iowa.

ILLINOIS DISTRIBUTION: Scattered in the northern half of the state; also St. Clair County.

This taxon, which flowers from July to September, often grows with *Ammophila breviligulata*. Although Hitchcock and others would maintain these species are closely related, Reeder and Ellington (1960) suggest that *Calamovilfa* is very distinct in characters of the embryo, epidermal hairs, lodicules, leaf anatomy, and chromosomes.

Thieret (1960) gives evidence why var. *magna* should be recognized, pointing out that typical *C. longifolia* is smaller in all respects and more often has glabrous sheaths. The typical variety, which does not apparently occur in Illinois, ranges to the northwest of this state.

65. Muhlenbergia SCHREB. – Muhly

Perennials (in Illinois) with or without rhizomes; blades flat or involute; inflorescence a spike-like or open panicle; spikelets

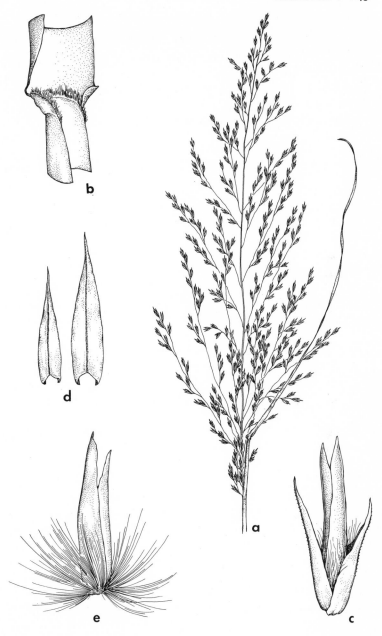

226. *Calamovilfa longifolia* var. *magna* (Sand Reed). *a.* Inflorescence, X½.
b. Sheath, with ligule, X5. *c.* Spikelet, X 7½. *d.* Glumes, X6. *e.* Lemma
and palea, X7½.

1-flowered, disarticulating above the glumes; glumes subequal, awned or awnless, keeled or rounded on the back; lemma rounded on the back, obscurely 3- to 5-nerved, often awned; fruit enclosed by the lemma.

Species of *Muhlenbergia* in the western United States are important range grasses.

For a detailed account of species 8, 9, 11, 13, and 14, see Fernald (1943).

Pohl (1969) has prepared an excellent treatment of species 4–14.

KEY TO THE SPECIES OF Muhlenbergia IN ILLINOIS

1. Panicle diffuse, open, at least 5 cm across; spikelets on pedicels longer than the lemmas.
   2. Rhizomes present; panicle 5–20 cm long; spikelets about 1.5 mm long; lemma glabrous, 1.3–1.7 mm long, awnless_____ _____1. *M. asperifolia*
   2. Rhizome lacking; panicle 20–45 cm long; spikelets (excluding awns) 3.0–4.5 mm long; lemma scabrous, 3.0–4.5 mm long, with an awn 5–20 m long_____2. *M. capillaris*
1. Panicle narrower, contracted to less than 2 cm thick.
   3. Plants densely tufted from firm bases or declined and rooting at the lower nodes, without rhizomes.
      4. Culms stiffly erect, tufted from firm bases; blades 1–2 mm broad, flat or involute; glumes 1.7–2.8 mm long; lemma acuminate, awnless_____3. *M. cuspidata*
      4. Lower portion of culms decumbent, rooting at the nodes; blades (1–) 2–4 mm broad, flat; glumes 0.1–1.5 (–1.7) mm long; lemma awned.
         5. Glumes obtuse, the first 0.1–0.2 mm long, the second 0.1–0.3 mm long; awn of lemma 1.5–4.0 mm long_____ _____4. *M. schreberi*
         5. Glumes acute to aristate, the first 0.5 mm long, the second 1.0–1.5 (–1.7) mm long; awn of lemma 0.5–1.5 mm long _____5. *M. × curtisetosa*
   3. Plants rhizomatous, not rooting at the lower nodes.
      6. Internodes glabrous or minutely scabrous, but not puberulent or pilose (*M. racemosa* rarely puberulent at summit of internodes).
         7. Panicle to 4 mm broad; glumes ovate-lanceolate, 1.3–2.0 (–2.5) mm long.
            8. Body of lemma 1.7–2.3 mm long, awnless or with an

awn 1–2 (–4) mm long_____6. *M. sobolifera*
8. Body of lemma 2.5–3.3 mm long, with an awn 2–7
mm long_____7. *M. bushii*
7. Panicle 5–15 mm broad; glumes lance-subulate, 1.6–8.0
mm long.
9. Spikelets 2–4 mm long; glumes 2–3 mm long_____
_____8. *M. frondosa*
9. Spikelets 4–8 mm long; glumes 4–8 mm long_____
_____9. *M. racemosa*
6. Internodes puberulent or pilose.
10. Lemma glabrous at base_____10. *M. glabrifloris*
10. Lemma pilose at base.
11. Glumes (including awns) (3.2–) 4.5–8.0 mm long
_____11. *M. glomerata*
11. Glumes 1.5–3.5 mm long.
12. Glumes ovate-lanceolate; anthers 1.0–1.5 mm
long; mature grain 2.0–2.3 mm long_____
_____12. *M. tenuiflora*
12. Glumes linear-lanceolate; anthers 0.3–0.8 mm
long; mature grain 1.3–1.8 mm long.
13. Glumes silvery or whitish; ligules 1.0–2.5
mm long; anthers 0.5–0.8 mm long_____
_____13. *M. sylvatica*
13. Glumes green or purplish; ligules 0.5–1.0
mm long; anthers 0.3–0.5 mm long_____
_____14. *M. mexicana*

1. **Muhlenbergia asperifolia** (Nees & Meyer) Parodi, Univ.

Nac. Buenos Aires Rev. Agron. 6:117. 1928. *Fig. 227.*

Vilfa asperifolia Nees & Meyer, Mem. Acad. St. Petersb. VI.
Sci. Nat. 4(1):95. 1840.
Sporobolus asperifolius Nees, Nov. Act. Acad. Caes. Leop.
Carol. 19: Sup. 1:9. 1841.
Cespitose perennial from scaly rhizomes; culms spreading or decumbent, to 50 cm long; sheaths more or less keeled; blades
1–2 mm broad, scabrous; panicle diffuse, spreading, to 20 cm
long; spikelets 1.0–1.7 mm long, purple; glumes acute, 1.0–1.5
mm long; lemma obtuse to subacute, glabrous, 1.3–1.7 mm long,
awnless.

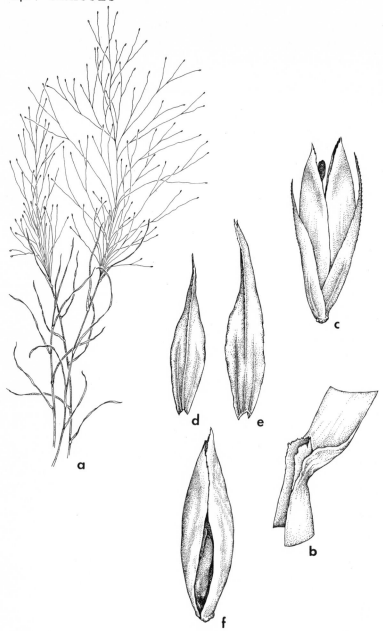

227. *Muhlenbergia asperifolia* (Scratch Grass). *a.* Upper part of plants, X½. *b.* Sheath, with ligule, X7½. *c.* Spikelet, X32½. *d.* First glume, X42½. *e.* Second glume, X42½. *f.* Lemma, X32½.

COMMON NAME: Scratch Grass.

HABITAT: Sandy soil.

RANGE: Minnesota to British Columbia, south to California, Texas, and Illinois; Mexico; South America.

ILLINOIS DISTRIBUTION: Known only from Champaign, Cook, DuPage, and Kane counties.

This species differs from all others of the genus in its open, diffuse panicles and small, glabrous spikelets. It flowers from June to September.

2. Muhlenbergia capillaris (Lam.) Trin. Gram. Unifl. 191. 1824. *Fig. 228.*

Stipa capillaris Lam. Tabl. Encycl. 1:158. 1791.

Podosaemum capillare (Lam.) Desv. Nouv. Bull. Soc. Philom. 2:188. 1810.

Cespitose perennial without rhizomes; culms erect, to nearly 1 m tall; sheaths glabrous or scabrous; blades involute, 2–4 mm broad, glabrous or scabrous; panicle diffuse, spreading, 20–45 cm long; spikelets (excluding the awns) 3.0–4.5 mm long; glumes ovate-lanceolate, acute or acuminate, the first 1.2–2.5 mm long, the second often aristate, 1.5–3.5 mm long; lemma narrow, acute, scabrous, 3.0–4.5 mm long, with an awn 5–20 mm long.

COMMON NAME: Hair Grass.

HABITAT: Sandy woodlands.

RANGE: Massachusetts to Kansas, south to Texas and Florida.

ILLINOIS DISTRIBUTION: Not common; confined to the southern counties.

This is one of the last species to begin to flower in Illinois, producing its flowers during September and October.

The longer, scabrous lemma, the large panicle, and the absence of rhizomes distinguish this delicate grass from *M. asperifolia.*

3. Muhlenbergia cuspidata (Torr.) Rydb. Bull. Torrey Club 32:599. 1905. *Fig. 229.*

Agrostis brevifolia Nutt. Gen. Pl. 1:44. 1818, non *M. brevifolia* Scribn. (1896).

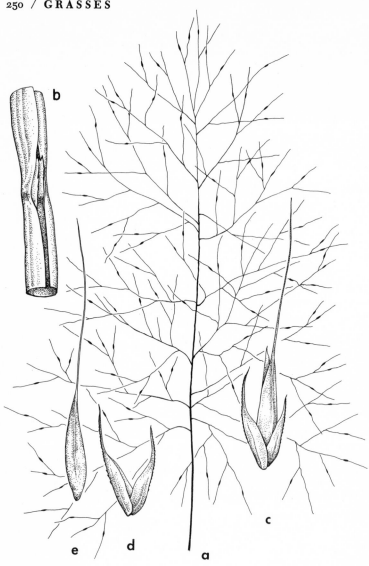

228. *Muhlenbergia capillaris* (Hair Grass). *a.* Inflorescence, X½. *b.* Sheath, with ligule, X5. *c.* Spikelet, X7½. *d.* Glumes, X7½. *e.* Lemma, X7½.

Vilfa cuspidata Torr. in Hook. Fl. Bor. Am. 2:238. 1840.
Sporobolus brevifolius (Nutt.) Scribn. Mem. Torrey Club 5:39. 1894.
Cespitose perennial, without rhizomes; culms strictly erect, to

229. *Muhlenbergia cuspidata* (Muhly). *a.* Upper part of plants, X½. *b.* Sheath, with ligule, 7½. *c.* Spikelet, X12½. *d.* First glume, X12½. *e.* Second glume, X12½.

75 cm tall; blades flat or involute, 1–2 mm broad; panicle slender, contracted, to 10 cm long; spikelets 3–4 mm long; glumes lance-subulate, 1.7–2.8 mm long; lemma narrower, acuminate, puberulent or nearly glabrous at base, awnless, 2.8–4.0 mm long.

COMMON NAME: Muhly.
HABITAT: Gravelly soil (in Illinois).
RANGE: Michigan to Alberta, south to New Mexico, Missouri, and Ohio.
ILLINOIS DISTRIBUTION: Rare; collected only in 1906 and 1912 by E. J. Hill from Will County (dry gravelly hills near Joliet, bank of DuPage River).
This species is distinct from other species of *Muhlenbergia* with contracted panicles by its firm culm-bases and absence of rhizomes. It flowers from mid-August to early October.

4. Muhlenbergia schreberi Gmel. Syst. Nat. 2:171. 1791. *Fig. 230.*

Muhlenbergia diffusa Willd. Sp. Pl. 1:320. 1797.
Dilepyrum minutiflorum Michx. Fl. Bor. Am. 1:40. 1803.
Muhlenbergia palustris Scribn. Bull. U.S.D.A. Div. Agrost. 11:47. 1898.

Perennial without rhizomes; culms decumbent below, rooting at the lower nodes, erect above, to 60 cm long; internodes glabrous; blades flat, 1–4 mm broad; panicle slender, contracted, to 10 cm long, to 6 mm broad; spikelets 1.5–2.5 mm long; glumes broadly ovate to orbicular, erose at the apex, the first 0.1–0.2 mm long, rarely absent, the second 0.1–0.3 mm long; lemma pubescent at base, 1.8–2.5 mm long, with an awn 1.5–4.0 mm long; n = 20 (Pohl, 1969).

COMMON NAME: Nimble Will.
HABITAT: Waste ground, woodlands.
RANGE: New Hampshire to Nebraska, south to Texas and Florida; Mexico.
ILLINOIS DISTRIBUTION: Common; in every county.
This species, a member of our native woodland flora, often becomes a troublesome weed in lawns. It begins to flower in early July, and persists until late October. Its minute glumes distinguished if from all other species of *Muhlenbergia*.
This species is the type of the genus *Muhlenbergia*.
The erect phase which is manifest in early spring is unlike the sprawling autumnal phase in appearance.
A specimen from Champaign County, collected by G. P.

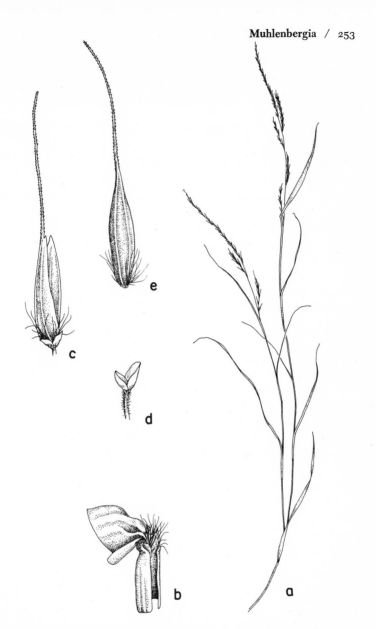

230. Muhlenbergia schreberi (Nimble Will). *a*. Upper part of plant, X½.
b. Sheath, with ligule, X5. *c*. Spikelet, X15. *d*. Glumes, X20. *e*. Lemma,
X15.

Clinton in 1892, is mentioned by Pohl (1969) as an apparent
hybrid with a rhizomatous species as one parent and *M. schreberi* as the other.

5. **Muhlenbergia** X **curtisetosa** (Scribn.) Pohl, Am. Midl. Nat. 82:528. 1969. *Fig. 231.*

Muhlenbergia schreberi ssp. *curtisetosa* Scribn. Rhodora 9:17. 1907.

Muhlenbergia curtisetosa (Scribn.) Bush, Am. Midl. Nat. 6:35. 1919.

Muhlenbergia schreberi var. *curtisetosa* (Scribn.) Steyerm. & Kucera, Rhodora 63:25. 1961.

Perennial, usually without rhizomes; culms decumbent below, rooting at the lower nodes, erect above, to 70 cm long; internodes glabrous; blades flat, 2–5 mm broad; panicle slender, contracted, to 10 cm long, to 7 mm broad; spikelets 2–3 mm long; first glume triangular, acute, 0.4–1.5 mm long; second glume ovate-lanceolate to ovate, acute to acuminate to aristate, 0.8–1.9 mm long; lemma pubescent at base, 2.2–3.4 mm long, with an awn 0.5–1.5 mm long.

HABITAT: Woodlands.

RANGE: Pennsylvania; Indiana; Illinois; Iowa; Missouri; Arkansas.

ILLINOIS DISTRIBUTION: Rare; known from Champaign and Fulton counties. The Champaign County specimen, collected by G. P. Clinton in 1892, has a short rhizome. The type was collected by J. Wolf in Illinois in 1881, presumably from Fulton County.

Since most pollen grains of this taxon are shrunken, Pohl (1969) suggests that *M.* X *curtisetosa* is a hybrid between *M. schreberi* and various rhizomatous species.

6. **Muhlenbergia sobolifera** (Muhl.) Trin. Gram. Unifl. 189. 1824.

Agrostis sobolifera Muhl. in Willd. Enum. Pl. 95. 1809.

Perennial from scaly rhizomes; culms erect, to nearly 1 m tall, the internodes glabrous or scaberulous; blades flat (2–) 4–7 mm broad, scabrous; panicle slender, contracted, to 15 cm long, to 4 mm broad; spikelets 1.7–2.7 mm long; glumes ovate-lanceolate, acuminate to cuspidate, 1.3–2.0 (–2.5) mm long; lemma acute or obtuse, mucronate, pilose at base, 1.7–2.3 mm long, awnless or with an awn up to 4 mm long; n = 20 (Pohl, 1969).

Two forms, based on presence or absence of a lemma awn, may be recognized in Illinois.

231. *Muhlenbergia* X *curtisetosa*. *a*. Upper part of plants, X½. *b*. Sheath, with ligule, X7½. *c*. Spikelet, X15.

1. Lemma awnless_____6a. *M. sobolifera* f. *sobolifera*
1. Lemma with an awn to 2 (-4) mm long_____
_____6b. *M. sobolifera* f. *setigera*

6a. Muhlenbergia sobolifera (Muhl.) Trin. f. **sobolifera** *Fig.* *232a–d.*

Lemma awnless.

COMMON NAME: Muhly.
HABITAT: Dry, rocky woodlands.
RANGE: New Hampshire to Wisconsin, south to Texas and Virginia.
ILLINOIS DISTRIBUTION: Occasional throughout the state. The ovate-lanceolate glumes relate this species to *M. tenuiflora*. *Muhlenbergia tenuiflora*, however, has puberulent internodes. Pohl (1969) reports a Johnson County specimen (*Pohl 9956*) which is unusually tall and leafy consistently had n = 21 at anaphase I.

6b. Muhlenbergia sobolifera (Muhl.) Trin. f. **setigera** (Scribn.) Deam, Publ. Ind. Dept. Conserv. 82:163. 1929. *Fig. 232e–f.*

Muhlenbergia sobolifera var. *setigera* Scribn. Rhodora 9:18. 1907.

Lemma with an awn to 2(–3) mm long.

HABITAT: Dry woodlands.
RANGE: Indiana and Illinois, south to Texas and Arkansas.
ILLINOIS DISTRIBUTION: Rare; only known from Wabash County.

7. Muhlenbergia bushii Pohl, Am. Midl. Nat. 82:534. 1969. *Fig. 233.*

Muhlenbergia brachyphylla Bush, Am. Midl. Nat. 6:41. 1919, non *M. brachyphylla* (Nees) Jackson (1895).

Perennial from scaly rhizomes; culms erect, to nearly 1 m tall, the internodes glabrous; blades 2–5 mm broad, scabrous; panicle slender, contracted, to 15 cm long; spikelets 2.5–3.3 mm long; glumes narrowly ovate, acute, awned, 1.4–1.7 (–2.5) mm long;

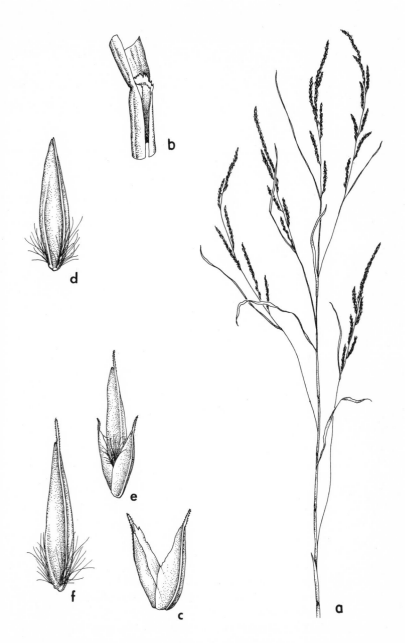

232. *Muhlenbergia sobolifera* (Muhly).—f. *sobolifera,* *a.* Upper part of plant, X½. *b.* Sheath, with ligule, X5. *c.* Glumes, X15. *d.* Lemma, X17½. —f. *setigera.* *e.* Spikelet, X15. *f.* Lemma, X17½.

lemma acuminate, pilose at the base, 2.5–3.3 mm long, the awn 2–7 mm long; 2n = 40 (Brown, 1950).

COMMON NAME: Muhly.

HABITAT: Mostly low woodlands.

RANGE: Indiana to Nebraska, south to Texas.

ILLINOIS DISTRIBUTION: Not common; in central and southern Illinois.

This is a rather rare species which perhaps is mistaken in the field for *M. tenuiflora* (with puberulent internodes) or *M. sobolifera* (with smaller lemmas and shorter or no awns). It flowers from August to October.

8. **Muhlenbergia frondosa** (Poir.) Fern. Rhodora 45:235. 1943.

Agrostis frondosa Poir. in Lam. Encycl. Sup. 1:252. 1790.

Agrostis lateriflora Michx. Fl. Bor. Am. 1:53. 1803.

Perennial from scaly rhizomes; culms sprawling or erect, to 1 m tall, the internodes glabrous; sheaths keeled; blades 2–7 mm broad; panicle slender, contracted to barely open, to 9 cm long, 5–9 mm broad; spikelets 2–4 mm long; glumes lance-subulate, 2–3 mm long; lemma acuminate, pilose at the base, 2.9–3.6 mm long, awnless or with the awn 4–11 mm long; n = 20 (Pohl, 1969).

Two forms occur in Illinois.

1. Lemma awnless_____8a. *M. frondosa* f. *frondosa*
1. Lemma with an awn 4–11 mm long_____
_____8b. *M. frondosa* f. *commutata*

8a. **Muhlenbergia frondosa** (Poir.) Fern. f. **frondosa** *Fig. 234a–e.*

Lemma awnless.

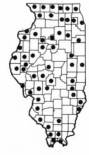

COMMON NAME: Muhly.

HABITAT: Moist woodlands; roadsides; fields.

RANGE: New Brunswick to North Dakota, south to Texas and Georgia.

ILLINOIS DISTRIBUTION: Rather common throughout the state.

For years this taxon has been called *M. mexicana,* a species with puberulent internodes. The flowering time for *M. frondosa* is August and September.

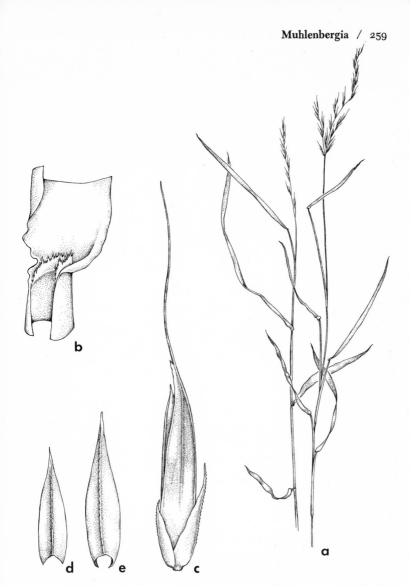

233. *Muhlenbergia bushii* (Muhly). *a.* Upper part of plants, X½. *b.* Sheath, with ligule, X5. *c.* Spikelet, X17½. *d.* First glume, X25. *e.* Second glume, X25.

8b. Muhlenbergia frondosa (Poir.) Fern. f. commutata

(Scribn.) Fern. Rhodora 45:235. 1943. *Fig. 234f–g.*

Muhlenbergia mexicana ssp. *commutata* Scribn. Rhodora 9:18. 1907.

Muhlenbergia umbrosa ssp. *attenuata* Scribn. Rhodora 9:21. 1907.

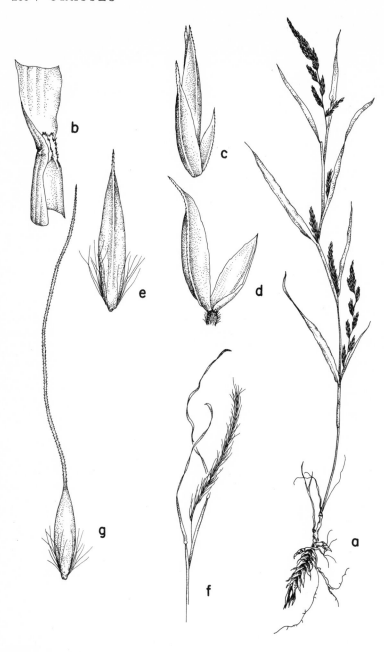

234. *Muhlenbergia frondosa* (Muhly),—*f. frondosa. a.* Habit, X½. *b.* Sheath, with ligule, X4. *c.* Spikelet, X12½. *d.* Glumes, X15. *e.* Lemma, K12½.—*f. commutata. f.* Inflorescence, X½. *g.* Lemma, X12½.

Muhlenbergia mexicana var. *commutata* (Scribn.) Farw. Rep. Mich. Acad. Sci. 17:181. 1916.

Muhlenbergia commutata (Scribn.) Bush, Am. Midl. Nat. 6:61. 1920.

Muhlenbergia mexicana f. *commutata* (Scribn.) Wieg. Rhodora 26:1. 1924.

Muhlenbergia umbrosa f. *attenuata* (Scribn.) Deam, Publ. Ind. Dept. Conserv. 82:171. 1929.

Muhlenbergia sylvatica f. *attenuata* (Scribn.) Palmer & Steyerm. Ann. Mo. Bot. Gard. 22:467. 1935.

Muhlenbergia diffusa var. *attenuata* (Scribn.) Farw. Papers Mich. Acad. Sci. 23:125. 1938.

Lemma with an awn 4–11 mm long.

HABITAT: Same as f. *frondosa*.

RANGE: Quebec to South Dakota, south to Missouri and Virginia.

ILLINOIS DISTRIBUTION: Occasional throughout the state, but not as common as f. *frondosa*.

9. **Muhlenbergia racemosa** (Michx.) BSP. Prel. Cat. N. Y. 67. 1888. *Fig. 235.*

Agrostis racemosa Michx. Fl. Bor. Am. 1:53. 1803.

Perennial from scaly rhizomes; culms erect, rarely declining, to 80 cm long, the internodes glabrous; sheaths keeled; ligule 0.6–1.5 mm long; blades 2–5 mm broad, scabrous; panicle narrow, contracted, to 17.5 cm long, 5–18 mm thick; spikelets 4–8 mm long (including the awns); glumes linear-subulate, scabrous, awned, 4–8 mm long; lemma acuminate, pilose at base, 2.5–4.0 mm long, awnless; n = 20 (Pohl, 1969).

235. *Muhlenbergia racemosa* (Marsh Muhly). *a.* Upper part of plants, X½.
b. Rhizome, X½. *c.* Sheath, with ligule, X5. *d.* Spikelet, X10. *e.* Glumes,
X10. *f.* Lemma, X10.

COMMON NAME: Marsh Muhly.

HABITAT: Dry soil.

RANGE: Newfoundland to British Columbia, south to Arizona, Oklahoma, and Virginia.

ILLINOIS DISTRIBUTION: Occasional; not common in the eastern counties.

This species resembles *M. glomerata* in the rather thick, contracted panicle. *Muhlenbergia glomerata*, however, has puberulent internodes.

Muhlenbergia racemosa flowers in August and September. The Michaux type is from southern Illinois.

10. Muhlenbergia glabrifloris Scribn. Rhodora 9:22. 1907.

Fig. 236.

Perennial from scaly rhizomes; culms erect, to 95 cm tall, the internodes puberulent near the apex; blades 2–4 (–7) mm broad; panicle slender, contracted, to 5 cm long; spikelets 2.2–3.0 mm long; glumes lance-subulate, acute or acuminate, scabrous, awnless, 2.2–3.0 mm long; lemma glabrous at base, scabrous near tip, 2.4–2.8 mm long, awnless; n = 20 (Pohl, 1969).

COMMON NAME: Muhly.

HABITAT: Moist woodlands.

RANGE: Maryland to Illinois, south to Texas and North Carolina.

ILLINOIS DISTRIBUTION: Occasional in the central and southern counties; absent from the northern counties. This species differs from all other rhizomatous species of *Muhlenbergia* with contracted panicles by its lemmas which are glabrous at the base.

11. Muhlenbergia glomerata (Willd.) Trin. Gram. Unifl. 191.

1824. *Fig. 237.*

Polypogon setosus Biehler, Pl. Nov. Herb. Spreng. Cent. 7. 1807, non *M. setosa* Kunth (1829).

Polypogon glomeratus Willd. Enum. Pl. 87. 1809.

Perennial from scaly rhizomes; culms more or less erect, to 120 cm tall, the internodes puberulent; sheaths scarcely keeled; ligule minute; blades 2–6 mm broad, scabrous; panicle narrow, contracted, to 11 cm long, 3–15 mm thick; spikelets 3.2–8.0 mm long; glumes linear-subulate, awned, 3.2–8.0 mm long; lemma

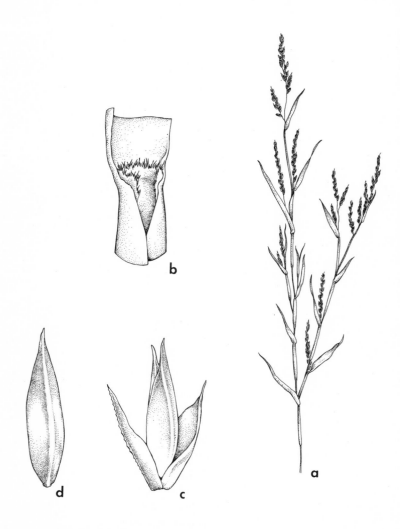

236. *Muhlenbergia glabrifloris* (Muhly). *a.* Upper part of plant, X½. *b.* Sheath, with ligule, X7½. *c.* Spikelet, X12½. *d.* Lemma, X12½.

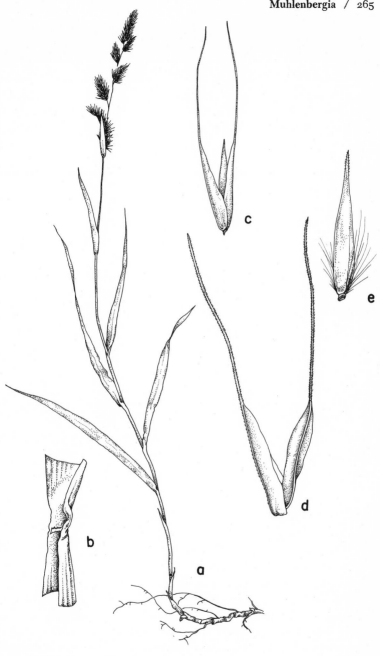

237. *Muhlenbergia glomerata* (Muhly). *a.* Habit, X½. *b.* Sheath, with ligule, X2½. *c.* Spikelet, X10. *d.* Glumes, X12½. *e.* Lemma, X12½.

pilose at base, acuminate, about 3 mm long, awnless; 2n = 40 (Avdulov, 1931); n = 10 (Pohl, 1969).

COMMON NAME: Muhly.

HABITAT: Dry or wet ground.

RANGE: Nova Scotia to Michigan, south to Illinois and Virginia.

ILLINOIS DISTRIBUTION: Rare; only three stations known in Illinois. (Lake: Illinois Beach State Park, *R. K. Brown 206;* St. Clair: Falling Spring, *H. Eggert s.n.;* also Cook County).

This species in the past has been confused with *M. racemosa,* a species with glabrous internodes, keeled sheaths, and very tiny anthers. *Muhlenbergia glomerata* last was collected in Illinois in 1951. It flowers during August and September. The St. Clair County specimen was collected from a bluff, a most peculiar habitat for this species.

12. Muhlenbergia tenuiflora (Willd.) BSP. Prel. Cat. N. Y. 67. 1888. *Fig. 238.*

Agrostis tenuiflora Willd. Sp. Pl. 1:364. 1797.

Muhlenbergia willdenovii Trin. Gram. Unifl. 188. 1824.

Perennial from scaly rhizomes; culms erect, to 1 m tall, the internodes retrorsely pubescent; blades flat, 5–10 (–15) mm broad; panicle slender, contracted, to 30 cm long; spikelets (excluding the awn) 3–5 mm long; glumes ovate-lanceolate, subulate, keeled, 1.5–3.0 mm long; lemma acuminate, pilose at the base, 2.5–3.5 mm long, with an awn 4–12 mm long; n = 20 (Pohl, 1969).

COMMON NAME: Slender Muhly.

HABITAT: Rocky woods; moist bluffs.

RANGE: Vermont to Wisconsin and Iowa, south to Missouri and Georgia.

ILLINOIS DISTRIBUTION: Occasional throughout the state. This is one of the more common species of *Muhlenbergia* which grows in rocky woodlands. Its puberulent internodes distinguish it from other similar species such as *M. sobolifera, M. bushii,* and *M. frondosa.* It may be differentiated from *M. sylvatica* and *M. mexicana* by its broader glumes. *Muhlenbergia tenuiflora* flowers from late July to early October.

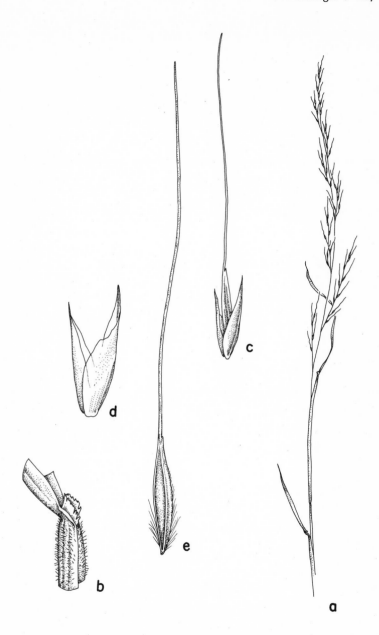

238. *Muhlenbergia tenuiflora* (Slender Muhly). *a.* Upper part of plant,
X1. *b.* Sheath, with ligule, X8. *c.* Spikelet, X10. *d.* Glumes, X15. *e.*
Lemma, X12½.

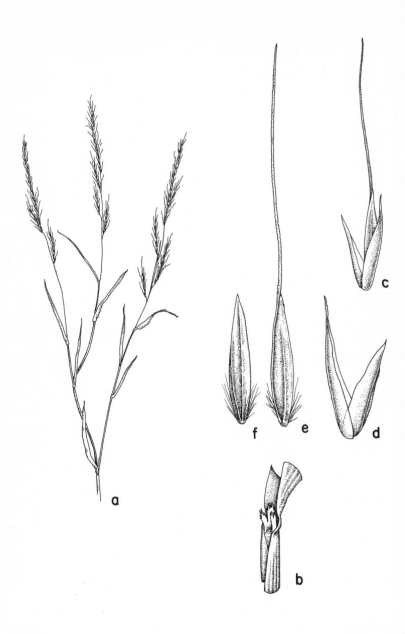

239. *Muhlenbergia sylvatica* (Muhly). *a.* Upper part of plant, X1. *b.* Sheath, with ligule, X8. *c.* Spikelet, X10. *d.* Glumes, X12½. *e.* Lemma, X12½. *f.* Lemma (awnless form), X12½.

13. **Muhlenbergia sylvatica** (Torr.) Torr. in Torr. & Gray, N. Am. Gram. & Cyp. 1:13. 1834. *Fig. 239.*

Agrostis diffusa Muhl. Descr. Gram. 64. 1817, non Host (1809).

Agrostis sylvatica Torr. Fl. North. & Mid. U. S. 1:87. 1823.

Muhlenbergia umbrosa Scribn. Rhodora 9:20. 1907.

Perennial from scaly rhizomes; culms erect, to nearly 1 m tall, the internodes usually puberulent; ligule 1.0–2.5 mm long; blades flat, 3–7 mm broad; panicle slender, contracted, erect or nodding, to 21 cm long; glumes linear-lanceolate, subulate, silvery or whitish, 1.8–3.0 mm long; lemma narrowed to the apex, pilose at the base, 2.2–3.5 mm long, with an awn 4–18 mm long, rarely awnless; 2n = 40 (Avdulov, 1931).

COMMON NAME: Muhly.

HABITAT: Rich, often rocky woods.

RANGE: Quebec to Ontario, south to Texas and Alabama.

ILLINOIS DISTRIBUTION: Occasional throughout the state. *Muhlenbergia sylvatica* may be distinguished from the rather common *M. mexicana* f. *mexicana* by the presence of an awn on the lemma of *M. sylvatica*. The awned *M. mexicana* f. *ambigua* is differentiated from *M. sylvatica* by its green spikelets, whereas the spikelets of *M. sylvatica* are silvery or whitish.

Muhlenbergia sylvatica flowers from early August to October.

14. **Muhlenbergia mexicana** (L.) Trin. Gram. Unifl. 189. 1824.

Agrostis mexicana L. Mant. Pl. 1:31. 1767.

Perennial from scaly rhizomes; culms erect, rarely decumbent, to nearly 1 m tall, the internodes puberulent; ligule 0.5–1.0 mm long; blades flat, 2–6 mm broad; panicle slender, contracted, to 21 cm long, 2–10 mm thick; spikelets often purplish, 2.5–3.5 mm long (excluding the awn); glumes lance-subulate, acute or aristate, 2.0–3.5 mm long; lemma acuminate, pilose at base, 2.2–3.5 mm long, awnless, or with an awn to 9 mm long; 2n = 40 (Avdulov, 1931).

Two forms occur in Illinois.

1. Lemma awnless_____14a. *M. mexicana* f. *mexicana*
1. Lemma with an awn to 9 mm long_____

_____14b. *M. mexicana* f. *ambigua*

14a. Muhlenbergia mexicana (L.) Trin. f. **mexicana** *Fig. 240a–f.*

Muhlenbergia mexicana β purpurea Wood, Am. Bot. & Flor. 2:386. 1870.

Muhlenbergia polystachya Mack. & Bush, Man. Fl. Jackson Co., Missouri 23. 1902.

Lemma awnless.

COMMON NAME: Muhly.

HABITAT: Moist soil, usually in woods.

RANGE: Quebec to British Columbia, south to California, Texas, and North Carolina.

ILLINOIS DISTRIBUTION: Occasional throughout the state. Plants with somewhat purplish spikelets, the type which was collected by Wolf from Illinois, have been known as var. *purpurea*, but this variation seems scarcely tenable. *Muhlenbergia mexicana* f. *mexicana* flowers from early August to mid-October.

14b. Muhlenbergia mexicana (L.) Trin. f. **ambigua** (Torr.)

Fern. Rhodora 45:236. 1943. *Fig. 240g–h.*

Agrostis filiformis Willd. Enum. Pl. 1:95. 1809, non Vill. (1787).

Agrostis lateriflora β filiformis (Willd.) Torr. Fl. North. & Mid. U. S. 1:86. 1823.

Muhlenbergia ambigua Torr. in Nicoll. Rep. Miss. 164. 1843.

Muhlenbergia foliosa ambigua Scribn. Rhodora 9:20. 1907.

Muhlenbergia foliosa f. *ambigua* (Torr.) Wieg. Rhodora 26:1. 1924.

Lemma with an awn to 9 mm long.

HABITAT: Same as f. *mexicana*.

RANGE: Quebec to South Dakota, south to Kansas and North Carolina.

ILLINOIS DISTRIBUTION: Not common; specimens seen from Lake County.

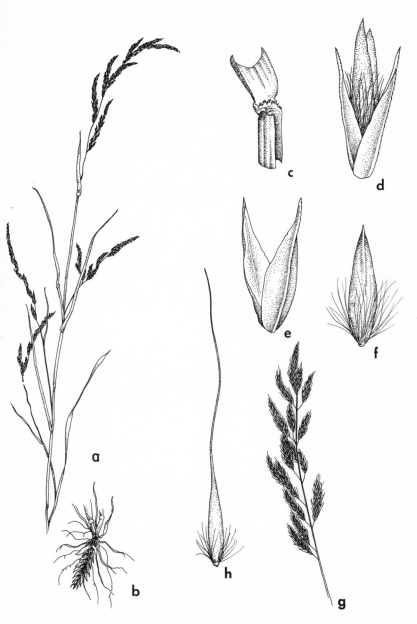

240. *Muhlenbergia mexicana* (Muhly),—f. *mexicana.*   *a.* Upper part of plant, X½.   *b.* Rhizome, X½.   *c.* Sheath, with ligule, X4.   *d.* Spikelet, X15. *e.* Glumes, X15.   *f.* Lemma, X12½.—f. *ambigua.*   *g.* Inflorescence, X½.   *h.* Lemma, X10.

66. *Sporobolus* R. BR. – Dropseed

Annuals or perennials; blades involute or flat; inflorescence paniculate, open or contracted and spike-like; spikelets 1-flowered, disarticulating above the glumes; glumes unequal or subequal, 1-nerved; lemma rounded on the back, more or less 1-nerved, awnless; palea at least as long as the lemma, sometimes broader than the lemma; grain quickly falling from the lemma and palea.

In western states, several species of *Sporobolus* are important forage grasses.

KEY TO THE SPECIES OF Sporobolus IN ILLINOIS

1. Panicle more or less open, spreading, over 10 cm long; perennials.
    2. Sheaths densely villous; spikelets 1.8–2.5 mm long; glumes acute, the first 1.0–1.5 mm long, the second 2.0–2.5 mm long; lemma 2.0–2.5 mm long_____1. *S. cryptandrus*
    2. Sheaths glabrous or nearly so; spikelets 4–6 mm long; glumes subulate or acuminate, the first 2–4 mm long, the second 4–6 mm long; lemma 3.5–5.5 mm long_____2. *S. heterolepis*
1. Panicle contracted, spike-like, less than 10 cm long (except in *S. asper*); perennials or annuals.
    3. Perennials; blades (before drying) 2.5 mm broad; panicle 5–25 cm long; only the upper sheaths swollen.
        4. Lemma glabrous; blade pilose on upper surface near base___
        _____3. *S. asper*
        4. Lemma sparsely villous; blade scabrous on margin and tip, but not pilose_____4. *S. clandestinus*
    3. Annuals; blades 1–2 mm broad; panicle 1–5 cm long; all or nearly all sheaths swollen.
        5. Spikelets 3.5–6.5 mm long; first glume 2.8–4.0 mm long; second glume 3.0–4.5 mm long; lemma 3–5 mm long, minutely villous_____5. *S. vaginiflorus*
        5. Spikelets 2–3 mm long; first glume 1.5–2.5 mm long; second glume 1.7–2.7 mm long; lemma 2–3 mm long, glabrous____
        _____6. *S. neglectus*

1. Sporobolus cryptandrus (Torr.) Gray, Man. 576. 1848.
Fig. 241.

Agrostis cryptandra Torr. Ann. Lyc. N. Y. 1:151. 1824.
Vilfa cryptandra Torr. ex Trin. Mem. Acad. St. Petersb. VI. Sci. Nat. 4(1):69. 1840.

Perennial with tufted or solitary culms to 1 m tall; sheaths densely villous; blades flat or becoming involute when dry, tapering to a filiform tip, scabrous on the margin, 2–6 mm broad; panicle

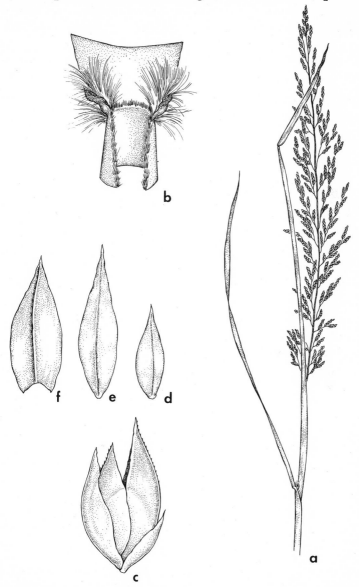

241. *Sporobolus cryptandrus* (Sand Dropseed). *a.* Inflorescence, X½. *b.* Sheath, with ligule, X5. *c.* Spikelet, X17½. *d.* First glume, X20. *e.* Second glume, X20. *f.* Lemma, X20.

ovoid to pyramidal, terminal and axillary, partly enclosed by the sheath at the base, more or less spreading, to 25 cm long; spikelets 1.8–2.5 mm long; glumes acute, glabrous, the first 1.0–1.5 mm long, the second 2.0–2.5 mm long, about equaling the lemma; lemma acute, glabrous, 2.0–2.5 mm long, equaling or longer than the palea; 2n = 18 (Nielsen, 1939), 36 (Brown, 1950).

COMMON NAME: Sand Dropseed.

HABITAT: Sandy soil.

RANGE: Quebec to Washington, south to Texas and North Carolina; Mexico.

ILLINOIS DISTRIBUTION: Occasional in the northern third of the state and in the western counties; apparently absent elsewhere.

The panicles are enclosed in the sheaths for varying lengths. The blades are, for the most part, flattened, but a few specimens with blades seemingly involute from the first have been seen. Sand Dropseed flowers from August to early October in Illinois.

2. Sporobolus heterolepis (Gray) Gray, Man. 576. 1848. *Fig. 242.*

Vilfa heterolepis Gray, Ann. Lyc. N. Y. 2:233. 1835.

Agrostis heterolepis (Gray) Wood, Class-book 598. 1847.

Tufted perennial with erect, wiry culms to 1 m tall; blades usually involute; panicle cylindric, spreading, purple to blackish, long-exserted from the sheath, to 25 cm long; spikelets 4–6 mm long; glumes glabrous, the first subulate, 2–4 mm long, the second acuminate, involute at the tip, 4–6 mm long; lemma glabrous, 3.5–5.5 mm long, slightly shorter than the palea.

COMMON NAME: Prairie Dropseed.

HABITAT: Dry soil, often on prairies.

RANGE: Quebec to Saskatchewan, south to Wyoming, Texas, and New York.

ILLINOIS DISTRIBUTION: Occasional in the northern half of the state; rare in the southern half.

This species, along with S. *cryptandrus,* are the only species of *Sporobolus* in Illinois with diffuse, more or less spreading, panicles. *Sporobolus heterolepis* flowers during August and September. It is one of the charac-

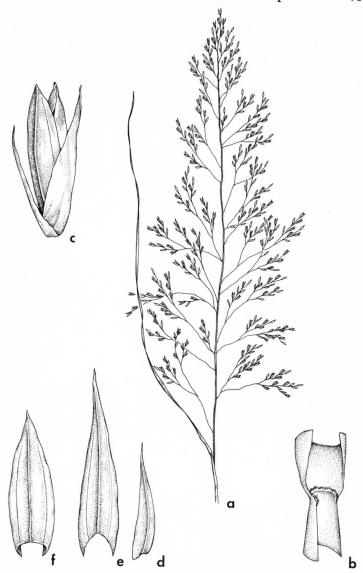

242. *Sporobolus heterolepis* (Prairie Dropseed). *a.* Inflorescence, X½. *b.* Sheath, with ligule, X5. *c.* Spikelet, X7½. *d.* First glume, X7½. *e.* Second glume, X7½. *f.* Lemma, X7½.

teristic grasses of the prairie, and is readily detected by its involute leaves.

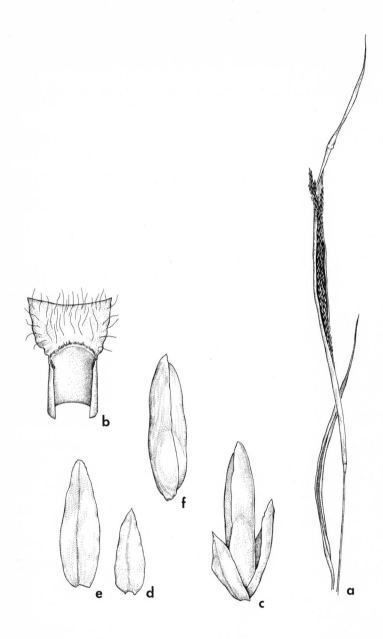

243. *Sporobolus asper* (Dropseed). *a*. Inflorescence, X½. *b*. Sheath, with ligule, X5. *c*. Spikelet, X7½. *d*. First glume, X7½. *e*. Second glume, X7½. *f*. Lemma, X7½.

3. **Sporobolus asper** (Michx.) Kunth, Rev. Gram. 1:68. 1829.
Fig. 243.

Agrostis aspera Michx. Fl. Bor. Amer. 1:52. 1803.
Vilfa aspera (Michx.) Beauv. Ess. Agrost. 16, 147, 181. 1812.
Vilfa drummondii Trin. Mem. Acad. St. Petersb. VI. Sci. Nat.
4(1):106. 1840.
Sporobolus drummondii (Trin.) Vasey, Descr. Cat. Grasses
U. S. 44. 1885.

Cespitose perennial with stout, erect culms to 1.2 m tall; blades involute when dry, tapering to a filiform tip, scabrous on the tip, pilose above near base, 2–4 mm broad; panicle pale or purplish, erect, contracted, to 15 (–25) cm long, mostly enclosed by the sheath; spikelets 5–7 mm long; glumes carinate, obtuse to acute, the first 2.0–3.5 mm long, the second 2.5–4.5 mm long; lemma carinate, glabrous, 3.5–6.0 mm long, as long as or slightly longer than the palea; 2n = 54, 108 (Brown, 1950).

COMMON NAME: Dropseed.
HABITAT: Dry, often sandy, soil.
RANGE: Quebec to North Dakota, south to Texas and Virginia.
ILLINOIS DISTRIBUTION: Occasional throughout the state. This is the most robust of the species of *Sporobolus* with contracted panicles. It flowers during September and October.

4. **Sporobolus clandestinus** (Biehler) Hitchc. Contr. U. S.
Nat. Herb. 12:150. 1908. *Fig. 244.*

Agrostis clandestina Biehler, Pl. Nov. Herb. Spreng. Cent. 8.
1807.
Sporobolus canovirens Nash in Britt. Man. 1042. 1901.
Sporobolus asper var. *canovirens* (Nash) Shinners, Rhodora
56:30. 1954.
Sporobolus clandestinus var. *canovirens* (Nash) Steyerm. &
Kucera, Rhodora 63:25. 1961.

Cespitose perennial with stout, erect culms to 1 m tall; blades involute (when dry), tapering to a filiform tip, scabrous on the margin and the tip, 2–5 mm broad; panicle pale, erect, contracted, more or less exserted from the sheath, to 10 cm long; spikelets 2–8 mm long; glumes lanceolate, acute to acuminate, the first 2–4 mm long, the second 2.8–5.2 mm long; lemma lanceolate, acute to acuminate, sparsely villous, 4.0–6.5 mm long, considerably shorter than to equaling the acute to acuminate palea.

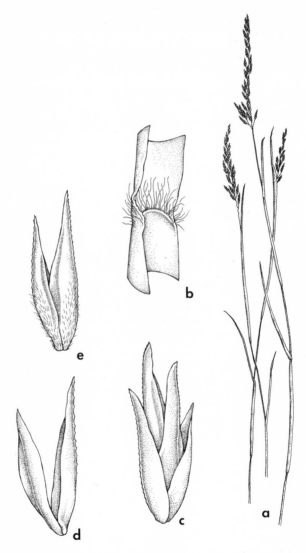

244. Sporobolus clandestinus (Dropseed). *a.* Upper part of plants, X½. *b.* Sheath, with ligule, X7½. *c.* Spikelet, X10. *d.* Glumes, X10. *e.* Lemma and palea, X10.

COMMON NAME: Dropseed.

HABITAT: Sandy soil.

RANGE: Connecticut to Kansas, south to Texas and Florida.

ILLINOIS DISTRIBUTION: Not common; restricted to a few counties.

This species flowers from late August to early October. Variation exists in the length of the palea. Specimens with the palea equaling the lemma have been designated *S. canovirens,* or *S. clandestinus* var. *canovirens,* while those with the palea exceeding the lemma by at least 2 mm have been considered "typical." I can see no reason for maintaining those with the shorter lemmas either as a species or as a variety.

5. **Sporobolus vaginiflorus** (Torr.) Wood, Class-book 775. 1861. *Fig. 245.*

Vilfa vaginiflora Torr. in Gray, N. Am. Gram. & Cyp. 1:3. 1834.

Sporobolus minor Vasey ex Gray, Man., ed. 6, 646. 1890.

Sporobolus vaginiflorus var. *inaequalis* Fern. Rhodora 35:109. 1933.

Annual with fibrous roots; culms cespitose, erect to spreading, wiry, scabrous, to 75 cm tall; sheaths ciliate at summit; blades involute at the tip, 1–2 mm broad, more or less glabrous; panicle slender, contracted, erect, more or less enclosed by the sheath, or the terminal becoming exserted at maturity, to 5 cm long; spikelets 3.5–6.5 mm long; glumes narrowly lanceolate, acuminate, the first 2.8–4.0 mm long, the second 3.0–4.5 mm long; lemma narrowly lanceolate, acuminate, minutely villous, 3–5 mm long, as long as or shorter than the palea.

COMMON NAME: Poverty Grass.

HABITAT: Dry soil.

RANGE: New Brunswick to North Dakota, south to Texas and Georgia.

ILLINOIS DISTRIBUTION: Occasional throughout the state. This species and *S. neglectus* are similar in their annual habit and their short, usually included, panicles. The spikelets of *S. vaginiflorus* are considerably larger than those of *S. neglectus.* Specimens with the paleas prolonged beyond the lemmas have been called var. *inaequalis*. I have been unable to convince myself that this is a justifiable segregation.

Sporobolus vaginiflorus flowers during September and October.

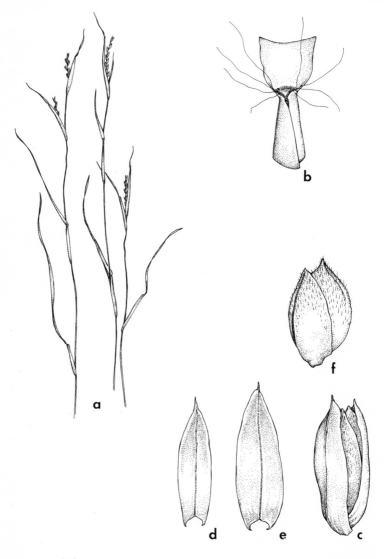

245. *Sporobolus vaginiflorus* (Poverty Grass). *a.* Upper part of plants, X½. *b.* Sheath, with ligule, X5. *c.* Spikelet, X10. *d.* First glume, X20. *e.* Second glume, X10. *f.* Lemma and palea, X10.

6. Sporobolus neglectus Nash, Bull. Torrey Club 22:464.
1895. *Fig. 246.*

Sporobolus vaginiflorus var. *neglectus* (Nash) Scribn. Bull.
U.S.D.A. Div. Agrost. 17:170. 1901.

Annual with fibrous roots; culms cespitose, erect to spreading,
wiry, to 75 cm tall; blades involute at the tip, pubescent near the
base, 1–2 mm broad; panicle slender, contracted, erect, often
purplish, mostly included in the sheath, to 5 cm long; spikelets
2–3 mm long; glumes narrowly lanceolate, acute to acuminate,
glabrous, the first 1.5–2.5 mm long, the second 1.7–2.7 mm long;
lemma narrowly lanceolate, acute to acuminate, glabrous, 2–3
mm long, about equaling the palea; 2n = 36 (Brown, 1950).

COMMON NAME: Sheathed Dropseed.
HABITAT: Dry soil.
RANGE: New Brunswick to North Dakota, south to Texas
and Virginia.
ILLINOIS DISTRIBUTION: Occasional throughout the state.
This species flowers during September and October. It
often grows in association with *S. vaginiflorus.*

67. Crypsis AIT.

Tufted annuals; blades more or less flat; panicles contracted,
spike-like, partly included in inflated sheaths; spikelets 1-flowered,
disarticulating above the glumes (in the Illinois species); glumes
subequal, keeled; lemma more or less keeled, 1-nerved; palea
nearly equaling the lemma; grain free from seed coat.

Lorch, who revised this genus in 1962, included the following
species, previously known as *Heleochloa schoenoides* (L.) Host,
in *Crypsis.*

1. Crypsis schoenoides (L.) Lam. Tabl. Encyc. 1:166, pl. 42.
1791. *Fig. 247.*

Phleum schoenoides L. Sp. Pl. 60. 1753.
Heleochloa schoenoides (L.) Host, Icon. Gram. Austr. 1:23.
1801.

Cespitose annual with erect or prostrate, glabrous culms to 40
cm tall; blades flat, becoming involute at the tip, 2–4 mm broad;

panicle contracted, spike-like, to 4 cm long, to 1 cm thick; spikelets 2.5–3.0 mm long; glumes ciliate on the keel, the first 1.5–2.0 mm long, the second 1.7–2.3 mm long; lemma cuspidate, 1.7–2.5 mm long; 2n = 36 (Avdulov, 1931).

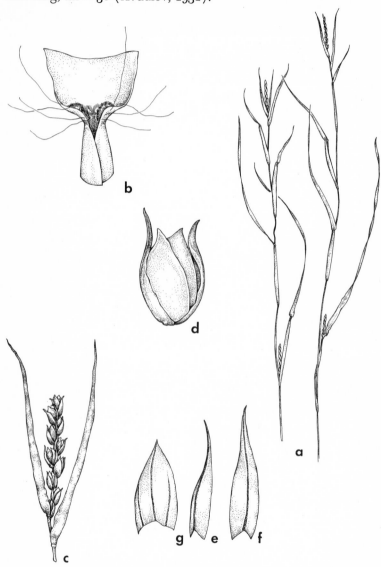

246. *Sporobolus neglectus* (Sheathed Dropseed). *a.* Upper part of plants, X½. *b.* Sheath, with ligule, X5. *c.* Inflorescence, X2. *d.* Spikelet, X10. *e.* First glume, X10. *f.* Second glume, X10. *g.* Lemma, X10.

HABITAT: Waste ground.

RANGE: Native of Europe; locally introduced in the United States.

ILLINOIS DISTRIBUTION: Known from several collections from Cook County and a single collection each from Grundy and St. Clair counties.

All Illinois collections were made in July, August, and September. The grain, which is free from the seed coat and is therefore not a caryopsis, is unique among most grasses.

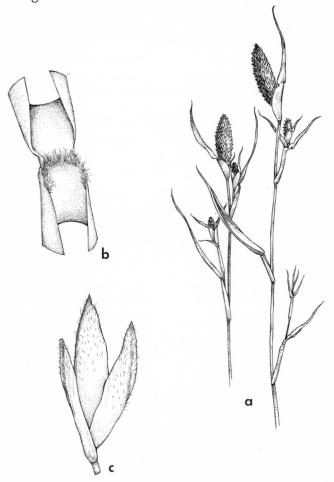

247. Crypsis schoenoides. a. Upper part of plants, X½. *b.* Sheath, with ligule, X5. *c.* Spikelet, X15.

Tribe *Chlorideae*

Spreading or tufted annuals or perennials; inflorescence composed of unilateral spikes, sometimes digitately arranged; spikelets 1- to several-flowered, sometimes unisexual; disarticulation above or below glumes.

Illinois genera placed in this tribe are *Eleusine, Dactyloctenium, Leptochloa, Gymnopogon, Schedonnardus, Cynodon, Chloris, Bouteloua, Buchloë,* and *Spartina.*

68. *Eleusine* GAERTN. – Goose Grass

Annuals; blades flat; inflorescence spicate, digitate, 1-sided; spikelets 3- to 6-flowered, disarticulating above the glumes; glumes unequal, strongly nerved, keeled; lemmas compressed, strongly nerved, keeled; palea shorter than the lemma, keeled.

Only the following species occurs in Illinois.

1. Eleusine indica (L.) Gaertn. Fruct. & Sem. 1:8. 1788.

Fig. 248.

Cynosurus indicus L. Sp. Pl. 72. 1753.

Annual; culms branched from near the base, spreading or ascending, to 50 cm long, compressed, glabrous; sheaths keeled, ciliate toward summit; blades flat or plicate, 3–8 mm broad; spikes (2–)3 3–8, to 10 cm long; spikelets appressed, 3- to 6-flowered; glumes scabrous on the keel, broadly lanceolate, subacute, the first 1–2 mm long, 1-nerved, the second 2–3 mm long, 3- to 5-nerved; lemmas lanceolate to ovate, subacute, 2.5–4.0 mm long; 2n = 18 (Avdulov, 1931), 36 (Moffett & Hurcombe, 1949).

COMMON NAME: Goose Grass.

HABITAT: Waste ground; troublesome in lawns.

RANGE: Native of Eurasia; naturalized in most parts of the United States.

ILLINOIS DISTRIBUTION: Common; probably in every county.

This troublesome weed flowers from late June to early October.

It resembles species of *Digitaria,* but the spikes are much thicker, the spikelets are more than 1-flowered, and the plants are generally much coarser.

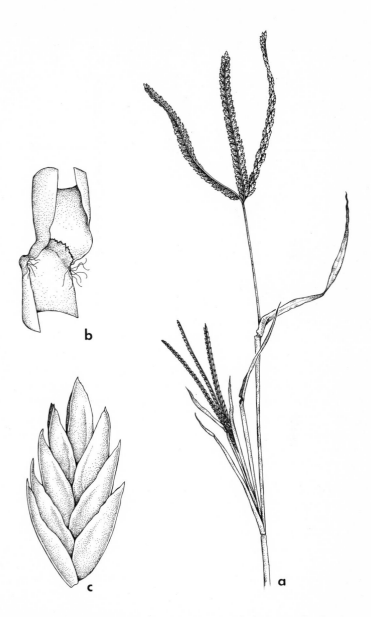

248. *Eleusine indica* (Goose Grass). *a.* Upper part of plant, X½. *b.* Sheath, with ligule, X5. *c.* Spikelet, X22½.

69. *Dactyloctenium* WILLD. – Crowfoot Grass

Annuals; blades flat; inflorescence spicate, digitate, 1-sided; spikelets 3- to 5-flowered, flattened, disarticulating above the first glume, not produced to the summit of the rachis; glumes sub-equal, 1-nerved, keeled, the second awned; lemmas flattened, keeled, usually short-awned, 3-nerved; palea as long as the lemma, keeled.

Only the following species occurs in Illinois.

1. Dactyloctenium aegyptium (L.) Beauv. Ess. Agrost. Expl. Pl. 15. 1812. *Fig. 249.*

Cynosurus aegyptius L. Sp. Pl. 72. 1753.

Annual; culms spreading, rooting at the nodes, glabrous, compressed, to 35 cm long; sheaths compressed, glabrous, overlapping; blades flat, ciliate along the margins, to 8 mm broad; spikes 2–5, to 5 cm long, the tips exposed and not spikelet-bearing; glumes ovate, scabrous on the keel, the first cuspidate, 1.0–1.5 mm long, the second 1.5–2.5 mm long, the awn 2–3 mm long, flexuous; lemmas broadly ovate, awn-tipped, 2.5–3.0 mm long; 2n = 36 (Moffett & Hurcombe, 1949).

COMMON NAME: Crowfoot Grass.

HABITAT: Waste ground.

RANGE: Native of the Old World; naturalized in the southern United States; California.

ILLINOIS DISTRIBUTION: Rare and probably extinct in Illinois; collected only once, in St. Clair County in 1876, by H. Eggert. The Illinois collection was made in August.

The exposed rachis-tips of the spikes distinguish this from other digitate species.

70. *Leptochloa* BEAUV. – Sprangletop

Annuals or perennials; culms usually branched; blades usually flat, rarely involute; inflorescence paniculate, composed of usually 1-sided racemes; spikelets 2- to 10-flowered, disarticulating above the glumes; glumes unequal, somewhat keeled, 1-nerved; lemmas rounded on the back, 3-nerved, with or without a short awn; palea about as long as the lemma.

The last three species enumerated below have been placed in the genus *Diplachne* by some workers.

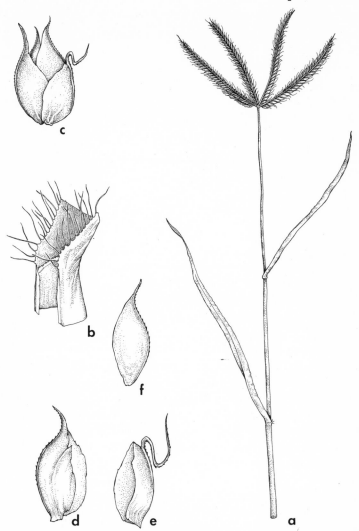

249. *Dactyloctenium aegyptium* (Crowfoot Grass). *a.* Upper part of plant, X½. *b.* Sheath, with ligule, X5. *c.* Spikelet, X12½. *d.* First glume, X12½. *e.* Second glume, X12½. *f.* Lemma, X12½.

KEY TO THE SPECIES OF Leptochloa IN ILLINOIS

1. Spikelets 2- to 4-flowered; lemmas 0.7–1.5 mm long; sheaths papillose-pilose; ligules 1–2 mm long.

    2. Lemmas 1.3–1.5 mm long; grain 0.7–0.9 mm long; glumes acute

--1. *L. filiformis*
2. Lemmas 0.7–1.0 mm long; grain 0.4–0.5 mm long; glumes aristate_____2. *L. attenuata*
1. Spikelets 5- to 10-flowered; lemmas 2–8 mm long; sheaths glabrous or nearly so; ligules 2–7 mm long.
   3. Blades usually becoming involute, 1–5 mm broad; lemma with an awn 0.5–1.0 mm long.
      4. Lemmas 2–4 mm long; panicle branches smooth or only slightly roughened_____3. *L. fascicularis*
      4. Lemmas (4–) 5–8 mm long; panicle branches very roughened_____4. *L. acuminata*
   3. Blades flat, 5–10 mm broad; lemma apiculate, awnless_____
--5. *L. panicoides*

1. **Leptochloa filiformis** (Lam.) Beauv. Ess. Agrost. 71, 161, 166. 1812. *Fig. 250.*

Festuca filiformis Lam. Tabl. Encycl. 1:191. 1791.
Eleusine mucronata Michx. Fl. Bor. Am. 1:65. 1803.
Leptochloa mucronata (Michx.) Kunth, Rev. Gram. 1:91. 1829.

Annual; culms simple or branched, to nearly 1 m tall; sheaths sparsely papillose-pilose; ligules 1–2 mm long; blades flat, scabrous on the margins, to 10 mm broad; panicle spreading, often purplish, to 50 cm long, the racemes 5–15 cm long; spikelets 2- to 3-flowered; glumes lanceolate, acute, the first 1.4–1.8 mm long, the second 1.5–2.0 mm long; lemmas lanceolate, subacute to obtuse, 1.3–1.5 mm long, pubescent on the nerves at the base; grain 0.7–0.9 mm long; 2n = 20 (Brown, 1950).

COMMON NAME: Red Sprangletop.
HABITAT: Low, sandy soil, particularly along rivers.
RANGE: Indiana to Kansas, south to New Mexico and Florida; Mexico; West Indies; South America.
ILLINOIS DISTRIBUTION: Occasional in the southern two-fifths of the state; absent elsewhere.
Synonymous with *L. filiformis* is *Eleusine mucronata*, the type of which was collected in Illinois by Michaux. *Leptochloa filiformis* flowers from late July to late September. It sometimes is the dominant grass along the banks of the major rivers in southern Illinois.

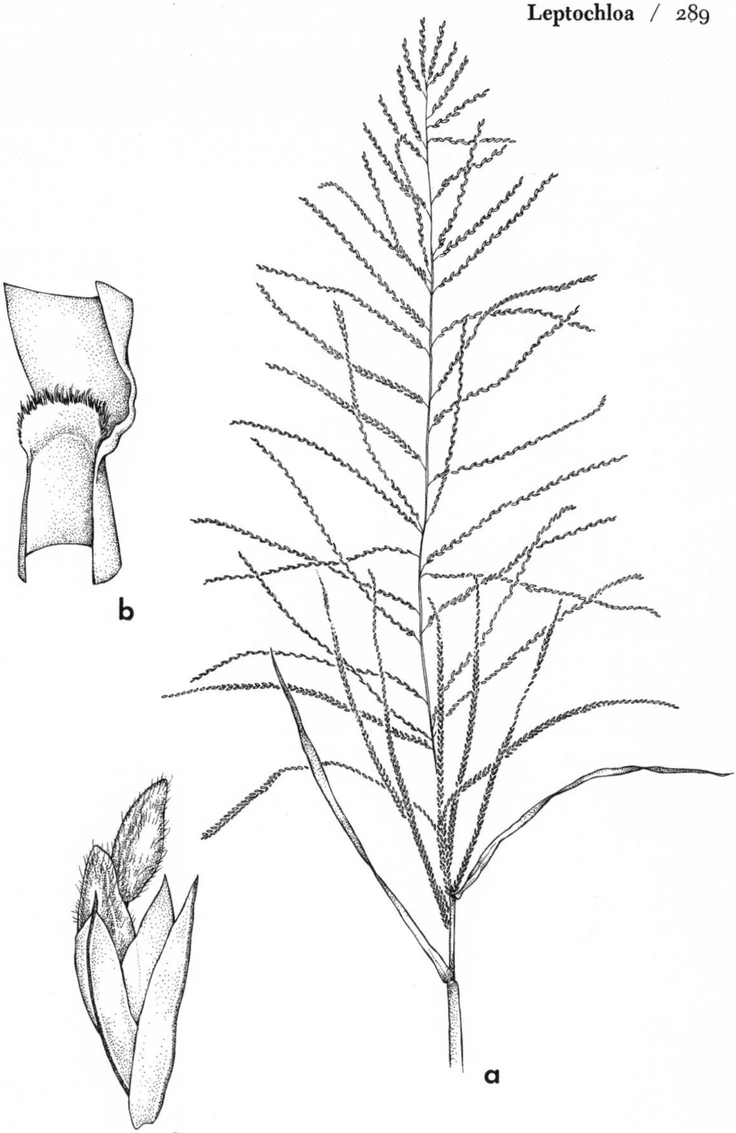

250. Leptochloa filiformis (Red Sprangletop). *a.* Inflorescence, X½. *b.* Sheath, with ligule, X5. *c.* Spikelet, X15.

2. Leptochloa attenuata (Nutt.) Steud. Syn. Pl. Glum. 1:209. 1854. *Fig. 251.*

Oxydenia attenuata Nutt. Gen. Pl. 1:76. 1818.

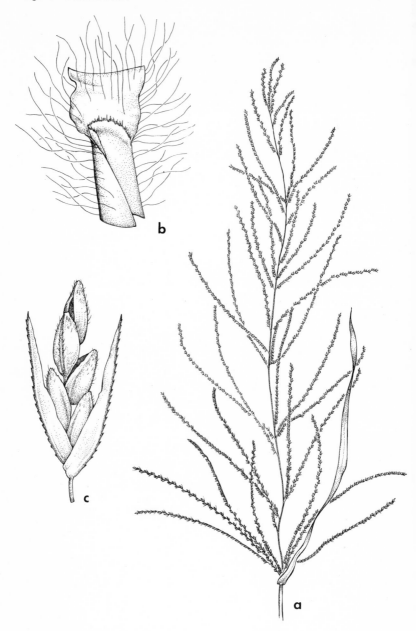

251. *Leptochloa attenuata* (Sprangletop). *a.* Inflorescence, X½. *b.* Sheath, with ligule, X5. *c.* Spikelet, X15.

Leptochloa filiformis f. *attenuata* (Nutt.) Gates, Contr. Kans. State Col. Agr. Dept. Bot. 391:130. 1940.

Leptochloa filiformis var. *attenuata* (Nutt.) Steyerm. & Kucera, Rhodora 63:26. 1961.

Annual; culms simple or branched, to nearly 1 m tall; sheaths sparsely papillose-pilose; ligules 1–2 mm long; blades flat, scabrous on the margins, to 10 mm broad; panicle spreading, to 35 cm long, the racemes to 11 cm long; spikelets 3- to 4-flowered; glumes narrowly lanceolate, aristate, variable in length from 1.3–2.5 mm; lemmas broadly lanceolate, broadly rounded at apex, 0.7–1.0 mm long, somewhat pubescent on the nerves at the base; grain 0.4–0.5 mm long.

COMMON NAME: Sprangletop.

HABITAT: Sandy shores.

RANGE: Illinois to Louisiana and Texas.

ILLINOIS DISTRIBUTION: Occasional in the southern tip of the state; absent elsewhere.

There is some doubt as to the specific distinctness of this species from *L. filiformis,* but on the basis of its shorter lemmas, shorter grains, and aristate glumes, *L. attenuata* seems to merit recognition as a species.

3. **Leptochloa fascicularis** (Lam.) Gray, Man. 588. 1848. *Fig. 252.*

Festuca fascicularis Lam. Tabl. Encycl. 1:189. 1791.

Festuca polystachya Michx. Fl. Bor. Am. 1:66. 1803.

Diplachne fascicularis (Lam.) Beauv. Ess. Agrost. 81. 1812.

Leptochloa polystachya (Michx.) Kunth, Rev. Gram. 1:91. 1829.

Tufted annual; culms densely branched from near the base, decumbent to erect, to nearly 1 m tall; sheaths more or less glabrous; ligules 5–7 mm long; blades flat or becoming involute, scabrous, 1–5 mm broad; panicle ascending, the base partly included in the sheath, to 35 cm long, the branches smooth or slightly scabrous; spikelets 5- to 10-flowered; glumes lanceolate, acute, scabrous on the keel, the first 1.3–2.0 mm long, the second 2.2–3.5 mm long; lemmas broadly lanceolate, obtuse to subacute, pubescent at the base and on the lateral nerves, 2–4 mm long, with a terminal awn 0.5–1.0 mm long.

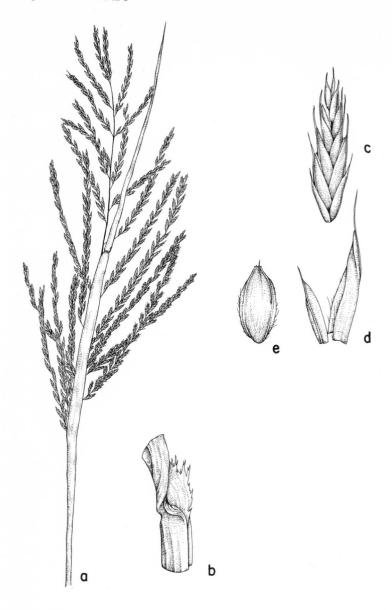

252. *Leptochloa fascicularis* (Salt Meadow Grass). *a.* Inflorescence, X½. *b.* Sheath, with ligule, X5. *c.* Spikelet, X6. *d.* Glumes, X12½. *e.* Lemma, X12½.

COMMON NAME: Salt Meadow Grass.

HABITAT: Wet soil.

RANGE: New Hampshire to Washington, south to California, Texas, and Florida; South America.

ILLINOIS DISTRIBUTION: Not common, but scattered throughout most of the state, except the northwestern counties and the extreme southern counties.

This and *L. acuminata* are the only species of *Leptochloa* in Illinois with awned lemmas. It differs from *L. acuminata* by its smaller lemmas and nearly smooth panicle branches.

It flowers from late July to mid-September.

Some authors would place this species in *Diplachne,* a genus not recognized in this work.

The type for Michaux's *Festuca polystachya* is from Illinois.

4. Leptochloa acuminata (Nash) Mohlenbrock, comb. nov.

Fig. 253.

Diplachne acuminata Nash in Britton, Man. 128. 1901.

Leptochloa fascicularis var. *acuminata* (Nash) Gl. Phytologia 4:21. 1952.

Tufted annual; culms branched from near the base, decumbent or occasionally ascending, to 8 dm tall; sheaths more or less glabrous; ligules 3–7 mm long; blades flat or becoming involute, harshly scabrous, to 5 mm broad; panicle ascending, partly included in the sheath, to 30 (–40) cm long, the branches scabrous; spikelets 5- to 10-flowered; glumes lanceolate, mostly aristate, scabrous on the keel, the first 2.0–3.7 mm long, the second 4–7 mm long; lemmas lanceolate, acuminate, pubescent on the lateral nerves (4–) 5–8 mm long, with a terminal awn 0.5–1.0 mm long.

COMMON NAME: Salt Meadow Grass.

HABITAT: Adventive along a railroad (in Illinois).

RANGE: Missouri to Colorado, south to Texas and Louisiana; adventive in Illinois.

ILLINOIS DISTRIBUTION: Kendall County (along CB & Q Railroad, in west sector of Yorkville, July 10, 1965, *F. A. Swink s.n.*).

Many authors do not segregate this predominantly western grass from *L. fascicularis.* It appears to me that the differences between the two, however, are clearly enough marked to warrant specific recognition.

253. Leptochloa acuminata (Salt Meadow Grass). *a.* Inflorescence, X ¼. *b.* Sheath, with ligule, X5. *c.* Spikelet, X6.

Leptochloa acuminata has much longer glumes and lemmas than *L. fascicularis,* as well as more harshly scabrous leaves and panicle branches.

254. *Leptochloa panicoides* (Salt Meadow Grass). *a.* Upper part of plant, X½. *b.* Sheath, with ligule, X2½. *c.* Lemma and palea, X10. *d.* Spikelet, X10.

5. **Leptochloa panicoides** (Presl) Hitchc. Am. Journ. Bot. 21:137. 1934. *Fig. 254.*

Megastachya panicoides Presl, Rel. Haenk. 1:283. 1830.
Poa panicoides (Presl) Kunth, Rev. Gram. 1:Sup. 28. 1830.

Diplachne halei Nash, Bull. N. Y. Bot. Gard. 1:292. 1899.
Leptochloa halei (Nash) Scribn. & Merr. Bull. U.S.D.A. Div.
Agrost. 24:27. 1901.
Annual; culms strong, branching, to nearly 1 m tall; sheaths more
or less glabrous; blades flat, scaberulous, 5–10 mm broad; panicle
rather lax, to 20 cm long, the branches ascending; spikelets 5- to
7-flowered; glumes ovate-lanceolate, acute, the first 1.0–1.8 mm
long, the second 1.5–2.2 mm long; lemmas ovate-lanceolate, obtuse to acute, apiculate, puberulent on the nerves at the base,
2.2–3.0 mm long, awnless.

COMMON NAME: Salt Meadow Grass.
HABITAT: Low areas.
RANGE: Indiana; Illinois; Missouri; Mississippi; Arkansas;
Texas; Brazil.
ILLINOIS DISTRIBUTION: Rare; known only from Pike and
Calhoun counties; first collected in Illinois in 1963 by
R. A. Evers.
Leptochloa panicoides is the only member of this genus
in Illinois with a 5- to 10-flowered spikelet and awnless
lemmas.
Its occurrence in west-central Illinois is far to the north of its
previously known range.

71. *Gymnopogon* BEAUV. – Beardgrass

Tufted rhizomatous perennials; blades short, broad, firm; spikes
numerous, slender, with remote, unilateral spikelets; spikelets
1-flowered and perfect (in the Illinois species) with the rachilla
projecting beyond the tip of the floret; disarticulation above the
glumes; glumes subequal, rigid, keeled, equaling or exceeding
the lemma; lemma awned from below the tip.
Only the following species occurs in Illinois.

1. **Gymnopogon ambiguus** (Michx.) BSP. Prel. Cat. N. Y. 69.
1888. *Fig. 255.*

Andropogon ambiguus Michx. Fl. Bor. Am. 1:58. 1803.
Tufted perennial from a short rhizome; culms ascending to erect,
stiff, to 60 cm tall; sheaths overlapping, glabrous; blades stiff,
crowded, spreading, to 60 mm long, to 15 mm broad, abruptly
pointed at the tip, rounded at the base; spikes to 20 cm long,
irregularly pinnately disposed and widely spreading along a cen-

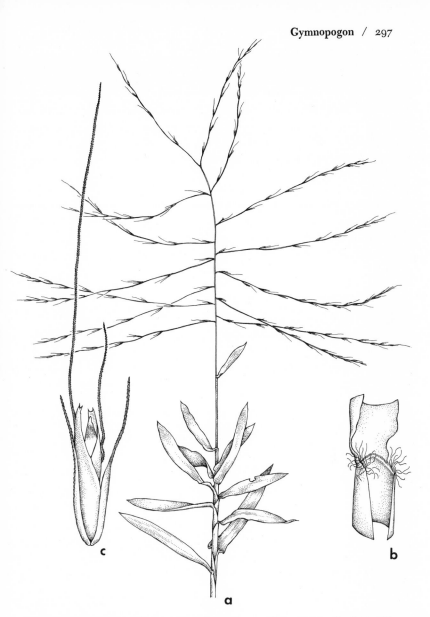

255. *Gymnopogon ambiguus* (Beardgrass). *a.* Upper part of plant, X½.
b. Sheath, with ligule, X4. *c.* Spikelet, X7½.

tral axis; spikelets remote, borne along one side of the rachis;
glumes narrow, 1-nerved, 4–6 mm long; lemma 3–6 mm long,
3-nerved, with an awn to 6 mm long; extension of rachilla (rudi-
ment) beyond floret to 3 mm long.

COMMON NAME: Beardgrass.

HABITAT: Sandy or gravelly soil in open areas.

RANGE: New Jersey to Kansas, south to Texas and Florida.

ILLINOIS DISTRIBUTION: Very rare; known only from Pope County (Burke Branch, November 16, 1966, *J. Schwegman, 1088*).

This is a highly distinctive grass because of the short and broad crowded leaves and the spreading, spicate panicle with remote spikelets.

Beardgrass is one of the rarest native grasses of Illinois, where it occurs in considerable abundance along Burke Branch, Pope County, in extreme southeastern Illinois.

72. *Schedonnardus* STEUD. – Tumble Grass

Tufted annual; blades flat; inflorescence composed of terminal and lateral spikes; spikelets 1-flowered, disarticulating above the glumes, disposed along two sides of the triangular rachis; glumes subequal, keeled, 1-nerved; lemma keeled, 3-nerved; palea much smaller than the lemma.

Only the following species comprises the genus.

1. **Schedonnardus paniculatus** (Nutt.) Trel. in Branner & Coville, Rep. Geol. Survey Ark. 1888 (4):236. 1891. *Fig. 256.*

Lepturus paniculatus Nutt. Gen. Pl. 1:81. 1818.

Tufted annual with erect or decumbent culms to 40 cm tall; blades mostly near the base, flexuous, 1–2 mm broad; inflorescence to 30 cm tall, composed of terminal and lateral spikes to 10 cm long; spikelets 3.5–4.5 mm long; glumes lance-subulate, scabrous on the keel, the first 2.0–3.2 mm long, the second 2.5–4.5 mm long; lemma narrowly lanceolate, acuminate, scabrous on the keel, 3.0–4.5 mm long; $2n = 30$ (Brown, 1950).

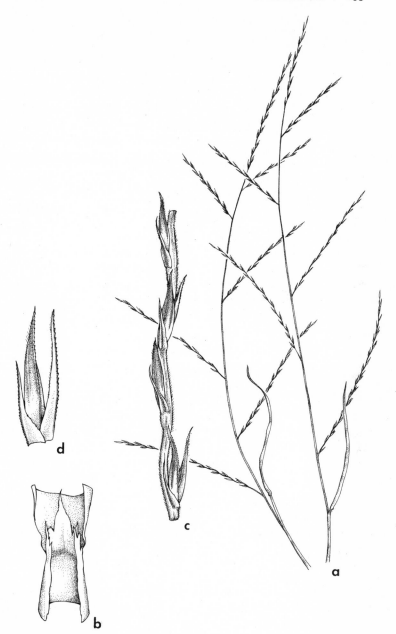

256. *Schedonnardus paniculatus* (Tumble Grass). *a.* Inflorescences, X½.
b. Sheath, with ligule, X4. *c.* Spike, X6. *d.* Spikelet, X9.

COMMON NAME: Tumble Grass.

HABITAT: Salt licks.

RANGE: Manitoba to Montana, south to New Mexico, Texas, and Illinois; Argentina.

ILLINOIS DISTRIBUTION: Very rare and no doubt extinct in Illinois; collected only twice, by Mead, from Hancock County, in 1845 and 1848.

The common name is derived from the inflorescence which often breaks away completely from the plant. The plant flowers during the summer.

73. *Cynodon* RICH. – Bermuda Grass

Rhizomatous perennials; blades flat; inflorescence spicate, digitate, 1-sided; spikelets 1-flowered, disarticulating above the glumes; glumes subequal, 1-nerved, keeled; lemma keeled, 3-nerved, awnless; palea about as long as the lemma.

Only the following species occurs in Illinois.

1. Cynodon dactylon (L.) Pers. Syn. Pl. 1:85. 1805. *Fig. 257.*

Panicum dactylon L. Sp. Pl. 58. 1753.

Digitaria dactylon (L.) Scop. Fl. Carn., ed. 2, 1:52. 1772.

Capriola dactylon (L.) Kuntze, Rev. Gen. Pl. 2:764. 1891.

Rhizomatous or stoloniferous, strongly creeping, perennial; culms flattened, wiry, glabrous, to 40 cm tall; blades flat, glabrous or pilose on the upper surface, to 4 mm broad; spikes 4–6, up to 7 cm long; glumes lanceolate, acute to cuspidate, 1.0–1.5 mm long; lemma flattened, broadly lanceolate, acute, 2.0–2.5 mm long, pubescent on the keel; 2n = 36 (Brown, 1950).

COMMON NAME: Bermuda Grass.

HABITAT: Waste ground, lawns.

RANGE: Native of Europe; naturalized in all but the north-central and northwestern United States.

ILLINOIS DISTRIBUTION: Occasional throughout the state, except for the extreme northern counties.

This species is an important pasture grass in the southern United States. It may serve as a tough, satisfactory lawn grass in southern Illinois. It flowers from July to October.

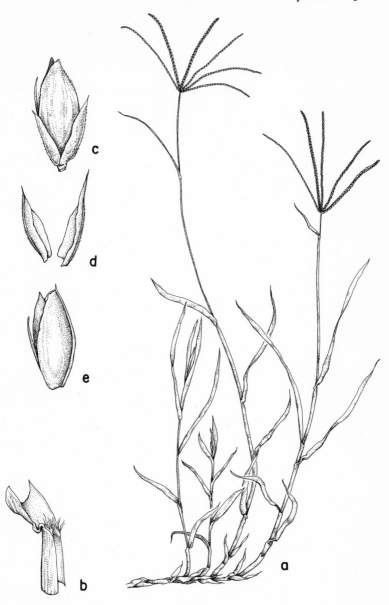

257. *Cynodon dactylon* (Bermuda Grass). *a*. Habit, X½. *b*. Sheath, with ligule, X4. *c*. Spikelet, X12½. *d*. Glumes, X15. *e*. Lemma and palea, X15.

74. *Chloris* SWARTZ – Finger Grass

Cespitose perennials; blades flat; inflorescence spicate, digitate; spikelets with 1 perfect flower and often several empty lemmas, borne on two sides of a triangular rachis, disarticulating above the glumes; glumes unequal, keeled; lemma keeled, 1- to 3-nerved, usually awned; palea nearly as long as the lemma.

Some species of the genus are valuable forage grasses in the southwestern United States.

KEY TO THE SPECIES OF Chloris IN ILLINOIS

1. Blades 1–3 mm broad, obtuse; awn of fertile lemma 5–9 mm long; empty lemma 1, with an awn 3.5–5.0 mm long; plants to 40 cm tall_____1. *C. verticillata*
1. Blades 3–5 mm broad, tapering to a long, fine point; awn of fertile lemma 1–5 mm long; empty lemmas 2, the upper awnless; plants usually over 1 m tall_____2. *C. gayana*

1. Chloris verticillata Nutt. Trans. Am. Phil. Soc. 5:150. 1837.

Fig. 258.

Tufted perennial; culms erect or decumbent, often rooting at the lower nodes, to 40 cm tall; blades 1–3 mm broad, obtuse; spikes widely spreading, to 15 cm long; spikelets about 3 mm long; glumes lance-subulate, keeled, the first 2–3 mm long, 1-nerved, the second 3–4 mm long, 1- to 3-nerved; fertile lemma acute, 3-nerved, pubescent on the nerves, 2.5–3.0 mm long, with the awn 5–9 mm long; empty lemma 1, about 1.5 mm long, with the awn 3.5–5.0 mm long; 2n = 80 (Brown, 1950).

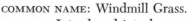

COMMON NAME: Windmill Grass.

HABITAT: Introduced into lawns in Illinois.

RANGE: Native to the western United States; introduced into several midwestern states.

ILLINOIS DISTRIBUTION: Known from Peoria, St. Clair, Tazewell, Winnebago, and Jackson counties. In the latter county, the species is abundant at several stations in Murphysboro.

The entire inflorescence often breaks away and tumbles before the wind. The species flowers from June to September in Illinois.

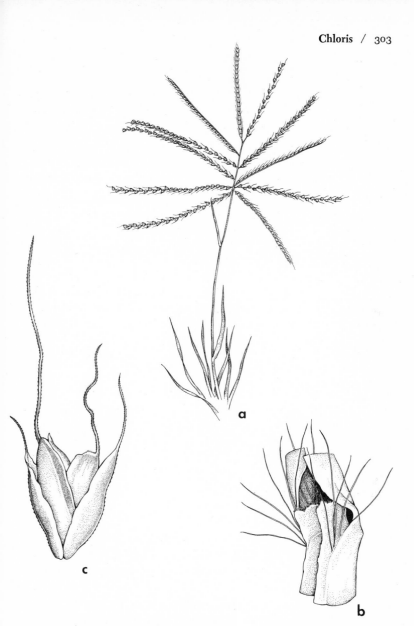

258. Chloris verticillata (Windmill Grass). *a.* Upper part of plant, X½.
b. Sheath, with ligule, X6. *c.* Spikelet, X10.

2. Chloris gayana Kunth, Rev. Gram. 1:89. 1829. *Fig. 259.*

Stoloniferous perennial; culms wiry, compressed, erect, usually at least 1 m tall; blades 3–5 mm broad, tapering to a fine point;

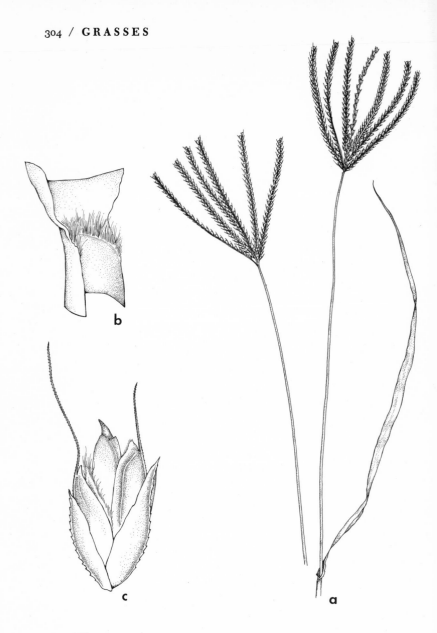

259. *Chloris gayana* (Finger Grass). *a*. Inflorescences, X½. *b*. Sheath, with ligule, X6. *c*. Spikelet, X12½.

spikes erect or ascending, 5–10 cm long; spikelets pale, 4–5 mm long; glumes acute to acuminate, the first 1–3 mm long, 1-nerved, the second 2.0–3.5 mm long, 1- to 3-nerved; fertile lemma acute, hispidulous, about 3 mm long, with the awn 1–5 mm long; empty lemmas 2, the uppermost truncate, awnless; 2n = 20 (Brown, 1950).

COMMON NAME: Finger Grass.
HABITAT: Introduced around a strip-mine in Illinois.
RANGE: Native of Africa.
ILLINOIS DISTRIBUTION: Known only from Perry County (strip-mine area, June 29, 1950, A. *Grandt s.n.*).

75. *Bouteloua* LAG. – Grama Grass

Perennials (in Illinois); blades flat or involute; inflorescence of short spikes borne along a common axis; spikelets with 1 perfect flower and 1–2 empty lemmas, disarticulating above the glumes, borne in 2 rows along one side of the rachis; glumes unequal, sometimes awn-tipped, 1-nerved; lemmas rounded on the back, usually 3-awned.

The genus is common in the western United States where most species are valuable forage grasses.

KEY TO THE SPECIES OF Bouteloua IN ILLINOIS

1. Spikes 10–50, spreading or nodding, to 2 cm long, falling entire at maturity; spikelets 7–10 mm long_____1. *B. curtipendula*
1. Spikes 1–3 (–6), straight or curved backward, 2–5 cm long, the florets falling from the glumes; spikelets 5–6 mm long.
   2. Rachis of inflorescence projecting 2–5 mm beyond last spike; second glume papillose-hirsute on keel; empty lemma glabrous at base; blades papillose-pubescent_____2. *B. hirsuta*
   2. Rachis of inflorescence not projecting; second glume villous on keel; empty lemma long-villous at base; blades glabrous or scabrous_____3. *B. gracilis*

1. Bouteloua curtipendula (Michx.) Torr. in Emory, Notes Mil. Recon. 154. 1848. *Fig. 260.*

Chloris curtipendula Michx. Fl. Bor. Am. 1:59. 1803.
Atheropogon apludoides Muhl. in Willd. Sp. Pl. 4:937. 1806.
Bouteloua curtipendula var. *aristosa* Gray, Man., ed. 2, 553. 1856.

Atheropogon curtipendulus (Michx.) Fourn. Mex. Pl. 2:138. 1886.

Bouteloua racemosa var. *aristosa* (Gray) Wats. & Coult. ex Gray, Man., ed. 6, 656. 1890.

Perennial from slender rhizomes; culms erect, to 1 m tall; sheaths pubescent near summit, otherwise more or less glabrous; blades flat or becoming involute at the tip, 2–5 mm broad, scabrous above and on the margins, glabrous or puberulent beneath; spikes 10–50, spreading or nodding, usually borne on one side of a common axis, to 2 cm long, falling entire; spikelets 3- to 6-flowered, 7–10 mm long; glumes subulate, the first linear, 3–4 mm long, the second lanceolate, 4–7 mm long, scabrous on the keel; fertile lemma acuminate, short-awned, scabrous, 4.5–8.0 mm long, with the two lateral nerves prolonged into awns about 1 mm long; empty lemma 1, with a long central awn and usually 2 shorter lateral ones; 2n = 28, 35, 40, 42, 45, 56, 70, 98 (Fults, 1942).

COMMON NAME: Side-oats Grama.

HABITAT: Prairies; dry hills.

RANGE: Ontario to Montana, south to Texas and Georgia; Mexico.

ILLINOIS DISTRIBUTION: Occasional in the northern half of the state; rather common on the bluffs bordering the Mississippi River from JoDaviess to Union counties. This species was originally collected by Michaux from Illinois. It differs from the other species of *Bouteloua* in Illinois in that its spikes fall entire. This character places it in Section Atheropogon.

Side-oats Grama is a conspicuous grass of the Illinois hill prairies. It flowers from mid-July to late September.

Specimens with the fertile lemma 3-awned, actually the common form in Illinois, have been described from Illinois material as var. *aristosa*.

2. Bouteloua hirsuta Lag. Var. Cienc. 4:141. 1805. *Fig. 261.*

Chondrosium hirtum HBK. Nov. Gen. & Sp. 1:176. 1816.
Atheropogon papillosus Engelm. Am. Jour. Sci. 46:104. 1843.
Chondrosium papillosum (Engelm.) Torr. in Marcy, Expl. Red Riv. 300. 1852.

Densely cespitose perennial; culms erect, to 60 cm tall; sheaths pilose at summit; blades mostly near base of plant, flat, 2–3 mm

260. *Bouteloua curtipendula* (Side-oats Grama). *a.* Inflorescences, X½. *b.* Sheath, with ligule, X6. *c.* Spikelet, X6. *d.* First glume, X7½. *e.* Second glume, X7½.

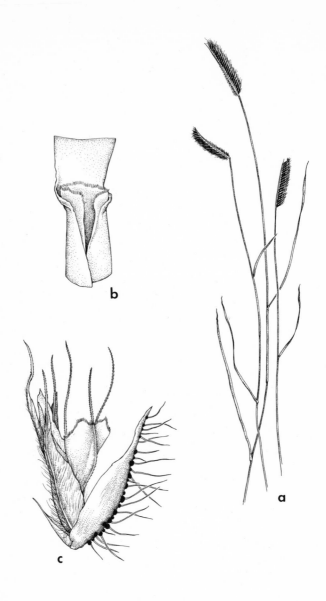

261. *Bouteloua hirsuta* (Grama Grass). *a.* Upper part of plants, X½.
b. Sheath, with ligule, X6. *c.* Spikelet, X12½.

broad, sparsely papillose-pubescent; spikes 1–3, straight or curved backward, 2–4 cm long, the rachis projecting at the tip for 2–5 mm; spikelets about 5 mm long; glumes 1-nerved, the first subulate, the second lanceolate, 3–4 cm long, papillose-hirsute on the keel; fertile lemma 3–4 mm long, sparsely villous, with 3 short, flat awns; empty lemma 1, stipitate, glabrous at the base, 3.5–5.0 mm long, 3-awned; 2n = 21, 37, 42 (Fults, 1942).

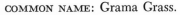

COMMON NAME: Grama Grass.
HABITAT: Prairies.
RANGE: Wisconsin to Colorado, south to Texas and Louisiana; Mexico.
ILLINOIS DISTRIBUTION: Not common; confined to the northwestern counties.
This and the following species belong to Section Chondrosium, with florets falling from the persistent glumes. *Bouteloua hirsuta* is distinguished from *B. gracilis* by its papillose-pubescent blades and glumes, its rachis which projects at the tip, and its empty lemma which is glabrous at the base. It flowers from mid-July to late September. A collection by Geyer from Cass County was described as *Atheropogon papillosus,* but this is identical with *B. hirsuta.*

3. **Bouteloua gracilis** (HBK.) Lag. ex Steud. Nom. Bot. 1:219. 1840. *Fig. 262.*

Chondrosium gracile HBK. Nov. Gen. & Sp. 1:176. 1816.
Atheropogon oligostachyus Nutt. Gen. Pl. 1:78. 1818.
Chondrosium oligostachyum (Nutt.) Torr. in Marcy, Expl. Red Riv. 300. 1852.
Bouteloua oligostachya (Nutt.) Torr. ex Gray, Man., ed. 2, 553. 1856.

Densely cespitose perennial from slender rhizomes; culms erect, to 60 cm tall; sheaths more or less glabrous; blades mostly near base of plant, flat or involute, curved or flexuous, 1–2 mm broad, glabrous or scabrous; spikes 1–3 (–6), straight or curved backward, 2–5 cm long, the rachis not projecting beyond the spikes; spikelets 5–6 mm long; glumes 1-nerved, the first subulate, the second lanceolate, 3.5–5.0 mm long, scabrous and villous on the keel; fertile lemma 3.5–5.0 mm long, densely pilose at base and on midnerve, with awns to 1.5 mm long; empty lemmas 2, the first stipitate, long-villous at the base, 3-awned, the second awnless; 2n = 20, 28, 35, 40, 42, 61, 62, 77, 84 (Fults, 1942; Snyder & Harlan, 1953).

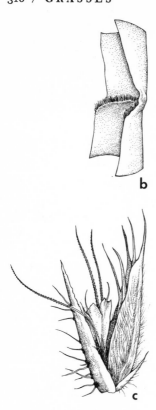

262. *Bouteloua gracilis* (Blue Grama). *a.* Upper part of plants, X½. *b.* Sheath, with ligule, X7½. *c.* Spikelet, X12½.

COMMON NAME: Blue Grama.

HABITAT: Sand flats.

RANGE: Manitoba to British Columbia, south to California, Texas, and Illinois; Mexico.

ILLINOIS DISTRIBUTION: Rare; known only from a collection made in 1908 from JoDaviess County (sand flats, July 18, 1908, *H. S. Pepoon 173*), and more recently from a strip-mine in Williamson County (July 19, 1950, *A. Grandt*).

76. Buchloë ENGELM. – Buffalo Grass

Dioecious perennials from stolons; blades flat, curly; inflorescence spicate, the staminate spikes 1–3, terminal, short, the pistillate spikes paired, head-like, mostly concealed among the

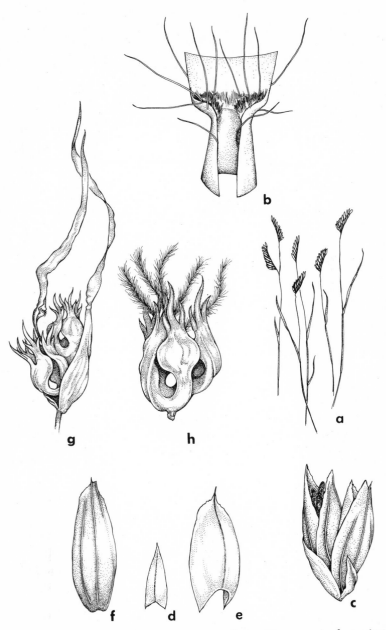

263. *Buchloë dactyloides* (Buffalo Grass). *a.* Upper part of staminate plants, X½. *b.* Sheath, with ligule, X7½. *c.* Staminate spikelet, X7½. *d.* First glume of staminate spikelet, X7½. *e.* Second glume of staminate spikelet, X7½. *f.* Lemma of staminate spikelet, X7½. *g.* Pistillate inflorescence, X3. *h.* Pistillate spikelet, X4½.

upper leaves; staminate spikelets 3-flowered, disarticulating above the glumes; glumes unequal, keeled, 1- to 3-nerved; lemmas keeled, 3-nerved; palea a little shorter than the lemma; pistillate spikelets 1-flowered; glumes unequal, the first often obsolete, the second indurate, rounded on the back, 3-lobed at the apex, the margins enclosing the lemma; lemma indurate, 3-lobed at the apex, enclosing the palea.

Only the following species comprises the genus.

1. Buchloë dactyloides (Nutt.) Engelm. Trans. Acad. Sci. St. Louis 1:432. 1859. *Fig. 263*.

Sesleria dactyloides Nutt. Gen. Pl. 1:65. 1818.

Stoloniferous perennials; culms bearing the staminate inflorescences to 30 cm tall; culms bearing the pistillate inflorescence very short, exceeded by the leaves; blades grayish-green, sparsely pilose, 1–2 mm broad; staminate spikes 5–15 mm long, elongated; pistillate spikes head-like, ovoid, 3–6 mm thick; 2n = 60 (Avdulov, 1931).

COMMON NAME: Buffalo Grass.

HABITAT: Prairies. This is one of the dominant grasses on the dry upland plains of the Great Plains.

RANGE: Manitoba to Montana, south to Texas and Louisiana; Mexico.

ILLINOIS DISTRIBUTION: Known only from a cemetery in Peoria County, collected there by V. H. Chase in 1956, and from a gravel bluff prairie in Winnebago County, collected there by E. W. Fell. The Illinois collections were made in August. This is the only dioecious species of tribe Chlorideae in Illinois.

Chase, who collected the first Illinois specimen, believes it is native since it grows in an undisturbed prairie remnant in a cemetery.

77. *Spartina* SCHREB. – Cord Grass

Perennials from scaly rhizomes; blades flat (in the Illinois species); inflorescence paniculate, composed of numerous 1-sided spikes; spikelets 1-flowered, disarticulating below the glumes; glumes unequal, keeled, awned, 1- to 3-nerved; lemma firm, keeled, 1- to 3-nerved; palea papery, 2-nerved, obscurely keeled, longer than the lemma.

For a monograph of the genus, see Mobberley (1956).

Only the following species occurs in Illinois.

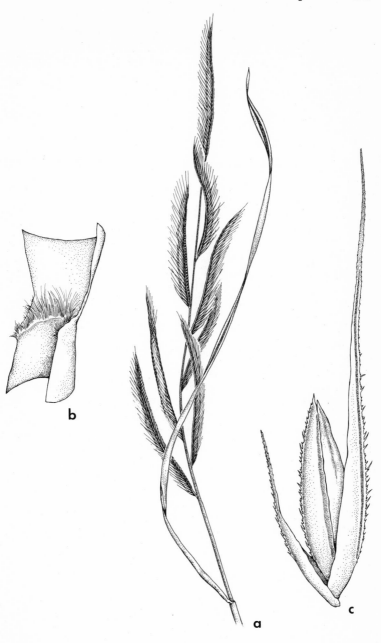

264. *Spartina pectinata* (Cord Grass). *a.* Upper part of plant, X½. *b.* Sheath, with ligule, X5. *c.* Spikelet, X6.

1. **Spartina pectinata** Link, Jahrb. Gewachs. I. 3:92. 1820.
Fig. 264.

Trachynotia cynosuroides Michx. Fl. Bor. Am. 1:64. 1803, non
Dactylis cynosuroides L. (1753).
Spartina michauxiana Hitchc. Contr. U. S. Nat. Herb. 12:153.
1908.

Perennial from rhizomes; culms to 2.5 m tall, erect, often solitary;
sheaths glabrous or pilose at the summit, the margins more or less
scabrous; blades flat, becoming involute when dry, to 15 mm
broad, scabrous on the margin, glabrous on both surfaces; panicle
to 50 cm long, composed of up to 50 mostly ascending spikes to
15 cm long; spikelets 10–25 mm long, appressed; glumes awned,
scabrous on the keel, the first 5–10 mm long, the second (including the awn) 10–25 mm long; lemma bilobed, apiculate, glabrous
except for the hispidulous margins and the more or less scabrous
keel, 7–10 mm long; 2n = 28 (Church, 1929).

COMMON NAME: Cord Grass.

HABITAT: Wet prairies; marshes.

RANGE: Newfoundland to Washington, south to Oregon,
Texas, and North Carolina.

ILLINOIS DISTRIBUTION: Occasional throughout the state,
except for the southern counties where it is rare.

This typical prairie species flowers from June to September.

Trachynotia cynosuroides was described from material
collected in Illinois.

The common name is derived from the tough scaly rhizomes.

Tribe *Zoysieae*

Although no member of tribe Zoysieae has as yet been found as
an escape in Illinois, the tribe is mentioned in this work because
of the increasing use of *Zoysia* as a lawn grass in the state.
Zoysia (*Fig. 102*) is distinguished by its 1-flowered, pedicellate
spikelets borne in zigzag spikes, its glabrous lemmas, its narrow
leaves, and its rhizomes or slender stolons.

Tribe *Aeluropodeae*

Stoloniferous or rhizomatous perennials; inflorescence a contracted panicle or raceme; spikelets 2- to several-flowered, often
unisexual; disarticulation above the glumes; lemmas several-
nerved.

The grasses which belong to this tribe usually inhabit saline or alkaline soils.

Only *Distichlis* represents tribe Aeluropodeae in Illinois.

78. *Distichlis* RAF. – Salt Grass

Dioecious perennials from creeping rhizomes; leaves involute; inflorescence small, dense, paniculate; spikelets 4- to 18-flowered, disarticulating above the glumes; glumes 2, unequal, several-nerved, keeled; lemmas firm, many-nerved, 2-ranked.

Only the following adventive species occurs in Illinois.

1. Distichlis stricta (Torr.) Rydb. Bull. Torrey Club 32:602. 1905. *Fig. 265.*

Uniola stricta Torr. Ann. Lyc. N. Y. 1:155. 1824.

Distichlis spicata stricta Scribn. Mem. Torrey Club 5:51. 1894. Perennial from slender, creeping rhizomes, with culms to 40 cm tall; blades rigid, mostly involute, serrulate, up to 10 cm long; inflorescence compact, 2–6 cm long; staminate spikelets 8- to 15-flowered, 12–25 mm long; pistillate spikelets 7- to 9-flowered, 10–17 mm long; glumes 3- to 7-nerved, keeled, acute, the first 3–6 mm long, the second 3.5–7.0 mm long; lemmas rigid, faintly 7- to 11-nerved, ovate, acute, 3.5–4.5 mm long; n = 20 (Stebbins & Löve, 1941).

COMMON NAME: Salt Grass.

HABITAT: Railroad right-of-way (in Illinois); otherwise, in alkaline soil.

RANGE: Minnesota to British Columbia, south to Missouri, Texas, and California; Mexico; adventive in Illinois.

ILLINOIS DISTRIBUTION: Rare; only two collections (Champaign Co.: Urbana, *H. E. Ahles 7400;* Cook Co.: *Thieret 2232*).

For a discussion of the variations of this species in North America, see Beetle (1943).

Tribe *Aristideae*

Annuals or perennials; blades narrow or involute; inflorescence paniculate; spikelets 1-flowered; disarticulation above the glumes; glumes large; lemma often 3-awned.

Embryo and anatomical characters have been responsible for the recent segregation of tribe Aristideae.

Only *Aristida* belongs to this tribe in Illinois.

265. *Distichlis stricta* (Salt Grass). *a.* Pistillate plant, X½. *b.* Staminate plant, X½. *c.* Sheath, with ligule, X6. *d.* Staminate spikelet, X5. *e.* First glume, X7½. *f.* Second glume, X7½. *g.* Lemma, X7½.

79. *Aristida* L. – Three Awn Grass

Tufted annuals or perennials; blades involute or flat and narrow; ligule obsolete; inflorescence paniculate or racemose, usually slender and narrow, equal or unequal, sometimes awned, the first often falling early; lemma indurate, very narrow, inrolled around the palea, with usually three awns.

Only in the southwestern United States is this genus valuable as a forage grass. None of the species in Illinois has any economic value.

A revision of the genus *Aristida* has been prepared by Henrard (1927).

KEY TO THE SPECIES OF Aristida IN ILLINOIS

1. Awns of lemma twisted and united into a basal column 8–15 mm long; glumes 20–30 mm long_____1. *A. tuberculosa*
1. Awns of lemma free to base or united into a basal column up to 3 mm long; glumes up to 20 mm long (except *A. oligantha* and *A. ramosissima*).
  2. Central awn of most lemmas at least 20 mm long.
    3. Lateral awns of lemma at least 12 mm long.
      4. Awns of lemma united at base into a column 1–3 mm long, the awns articulated with the summit of the lemma; sheaths villous on the margins_____2. *A. desmantha*
      4. Awns of lemma free to base, the awns not articulated with the summit of the lemma; sheaths glabrous or nearly so.
        5. All awns of lemma 35–70 mm long; glumes 12–32 mm long; lemma 12–20 mm long_____3. *A. oligantha*
        5. Central awn of lemma 15–33 mm long, the lateral awns 12–24 mm long; glumes 6–14 mm long; lemma 5.5–8.5 mm long.
          6. Perennial; first glume longer than second_____ _____4. *A. purpurascens*
          6. Annual; first glume equal to second or shorter.
            7. Glumes essentially equal in length; central awn of lemma longer than lateral awns_____ _____5. *A. intermedia*
            7. Second glume longer than first glume; awns of lemma equal in length_____6. *A. necopina*
    3. Lateral awns of lemma up to 6 mm long, or even absent____ _____7. *A. ramosissima*
  2. Central awn of lemma up to 20 mm long, usually considerably shorter.
    8. Lateral awns 12–20 mm long; central awn curved or divergent at base, not coiled.
      9. Central awn of lemma conspicuously bent near base; glumes obscurely 3-nerved_____8. *A. longespica*
      9. Awns of lemma not conspicuously bent near base; glumes sharply 1-nerved.

10. Glumes nearly equal in length; central awn of lemma longer than the lateral ones_____5. *A. intermedia*
10. Second glume longer than first glume; awns of lemma equal_____6. *A. necopina*
8. Lateral awns up to 12 mm long; central awn coiled at base (or bent conspicuously near base in *A. longespica*).
11. Glumes essentially equal, the second 4–10 mm long; lemma 4.0–8.5 mm long; inflorescence usually reduced to a raceme.
12. Central awn conspicuously bent at base, not coiled, 6.5–20.0 mm long; lateral awns 3–15 mm long; glumes obscurely 3-nerved_____8. *A. longespica*
12. Central awn loosely coiled at base, 3–10 mm long; lateral awns 0.7–3.3 mm long; glumes 1-nerved____ _____9. *A. dichotoma*
11. Second glume longer than first, usually by at least 2 mm, the second 7–15 mm long; lemma 7.5–10.5 mm long; inflorescence a slender panicle.
13. Lateral awns of lemma 5–12 mm long; central awn 10–19 mm long_____10. *A. basiramea*
13. Lateral awns of lemma 2–4 mm long; central awn 7.0–12.5 mm long_____11. *A. curtissii*

1. Aristida tuberculosa Nutt. Gen. Pl. 1:57. 1818. *Fig. 266.*

Annual; culms branched from near the base, erect, to 80 cm tall; lower sheaths villous, the upper more or less glabrous; blades involute, 1–3 mm broad; panicle lax, open, sparsely branched, to 30 cm long; glumes nearly equal, short-awned, 20–30 mm long, the first scabrous on the keel; lemma 10–15 mm long; awns equal, divergent, twisted and united at base into a column 8–15 mm long, the free parts loosely coiled, 30–50 mm long.

COMMON NAME: Needle Grass.
HABITAT: Sandy soil.
RANGE: Massachusetts to Georgia and Mississippi; Wisconsin; Minnesota; Indiana; Illinois; Iowa.
ILLINOIS DISTRIBUTION: Occasional in the northern half of the state; also Union County in 1884.
This species is related to *A. desmantha*, both belonging to Section Arthratherum, characterized by the awns of the lemma united below into a column. *Aristida tuberculosa* flowers from August to early October.

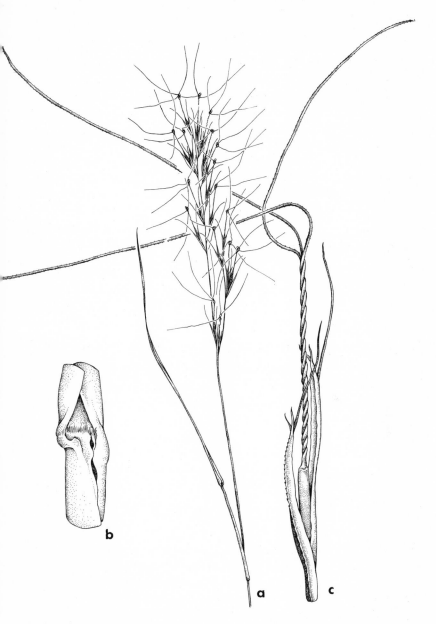

266. *Aristida tuberculosa* (Needle Grass).   *a.* Upper part of plant, X½.
b. Sheath, with ligule, X5.   *c.* Spikelet, X5.

2. **Aristida desmantha** Trin. & Rupr. Mem. Acad. St. Petersb. VI. Sci. Nat. 5(1):109. 1842. *Fig. 267.*

Annual; culms branched from near the base, erect, to 80 cm tall; sheaths villous on the margins; blades plicate or involute, 1–4 mm broad; panicle loosely branched, to 20 cm long, the branches ascending; glumes equal, short-awned, 13–20 mm long, the first scabrous on the keel; lemma 9–12 mm long; awns equal, loosely coiled near base and barely united into a column 1–3 mm long, the free parts 25–35 mm long.

COMMON NAME: Three Awn.

HABITAT: Sandy soil.

RANGE: Illinois; Nebraska; Texas.

ILLINOIS DISTRIBUTION: Very rare; known only from Cass, Mason, and Morgan counties.

This rare species has a disjunct distribution from three states. It was first found in Illinois in 1861. Although related to *A. tuberculosa*, it is differentiated clearly by its short column of awns. The Illinois specimens were collected in August and September.

3. **Aristida oligantha** Michx. Fl. Bor. Am. 1:41. 1803. *Fig. 268.*

Tufted annual; culms branched from near the base and at all nodes, wiry, to 50 cm tall; blades flat or involute, about 1 mm broad; panicle lax, few-flowered, to 20 cm long; glumes awned, the first 12–29 mm long, 3- to 5-nerved, scabrous on the keel, the second 14–32 mm long, 1-nerved, glabrous; lemma 12–20 mm long, scabrous near tip; awns equal, divergent, 35–70 mm long.

COMMON NAME: Three Awn.

HABITAT: Dry soil of fields and woods; along railroads.

RANGE: New York to South Dakota, south to Texas and Florida; California; Oregon.

ILLINOIS DISTRIBUTION: Common throughout the state. This is the most common species of *Aristida* in Illinois. It flowers from August to mid-October. It belongs to Section Chaetaria (with awns of lemma not united at base) and has the longest awns of any member of that section.

The type was collected by Michaux in Illinois.

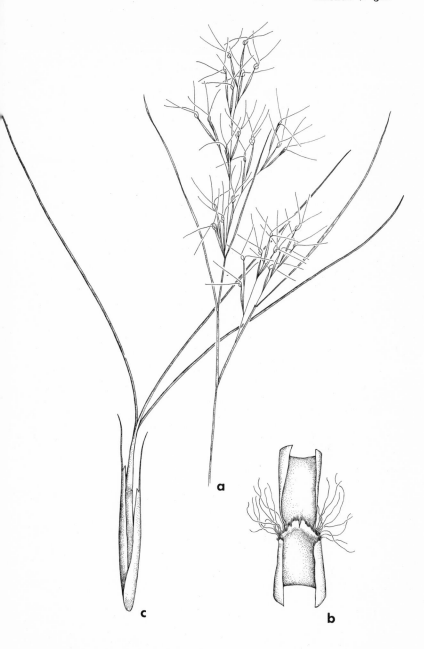

267. *Aristida desmantha* (Three Awn). *a.* Inflorescence, X½. *b.* Sheath, with ligule, X5. *c.* Spikelet, X3½.

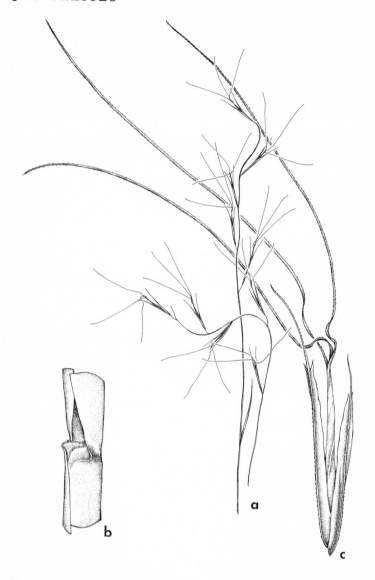

268. *Aristida oligantha* (Three Awn). *a.* Inflorescence, X½. *b.* Sheath, with ligule, X5. *c.* Spikelet, X4.

4. Aristida purpurascens Poir. in Lam. Encycl. Sup. 1:452. 1810. *Fig. 269.*

Aristida geyeriana Steud. Syn. Pl. Glum. 1:133. 1854.

Aristida stricta Steud. Syn. Pl. Glum. 1:133. 1854, non Michx. (1803).

Perennial from a knotty base; culms cespitose, sparsely branched, to 75 cm tall; sheaths glabrous or villous; blades 1–2 mm broad; panicle slender, spike-like, purplish, to 30 cm long, the branches ascending or appressed; glumes 1-nerved, aristate, the first 8–14 mm long, scabrous on the keel, the second 6.5–12.0 mm long; lemma 5.5–8.5 mm long; awns divergent, the central one 20–33 mm long, the lateral ones 15–24 mm long.

COMMON NAME: Arrowfeather.

HABITAT: Sandy soil.

RANGE: Massachusetts to Wisconsin, south to Kansas, Texas, and Florida.

ILLINOIS DISTRIBUTION: Occasional throughout the state, although not common in the southern counties.

This is the only perennial species of *Aristida* in Illinois. It flowers during August and September.

Steudel's *A. geyeriana*, obviously the same as *A. purpurascens*, originally was collected from Illinois.

5. **Aristida intermedia** Scribn. & Ball, Bull. U.S.D.A. Div. Agrost. 24:44. 1901. *Fig. 270.*

Annual; culms loosely cespitose, branched near the base, to 75 cm tall; blades flat or involute, scabrous, 1–2 mm broad; panicle loosely spike-like, very slender, to 30 cm long; glumes nearly equal, short-awned, scabrous on the keel, 1-nerved, 6.0–9.5 mm long; lemma 5.8–8.3 mm long, scabrous above the middle; awns divergent, the central one 15–24 mm long, the lateral ones 12–18 mm long.

COMMON NAME: Three Awn.

HABITAT: Sandy soil.

RANGE: Michigan to Nebraska, south to Texas and Mississippi.

ILLINOIS DISTRIBUTION: Rare; Henry County where first collected in 1939; also Lake, Lee, and McHenry counties.

This species resembles *A. purpurascens*, except for the annual habit and the shorter awns. It is also close to *A. longespica* var. *geniculata*, but this latter taxon has the central awn conspicuously bent near the base. It differs from

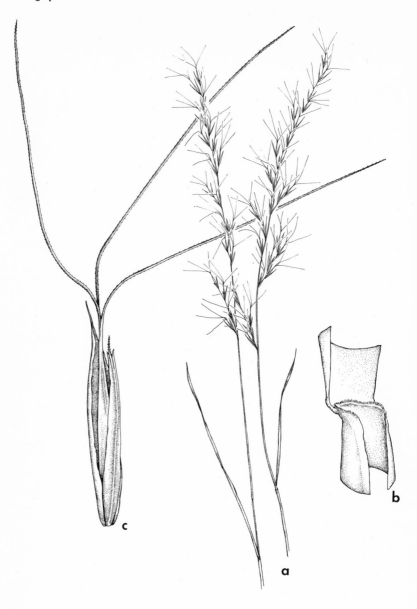

269. *Aristida purpurascens* (Arrowfeather).   *a*. Upper part of plants, X½.
b. Sheath, with ligule, X6.   *c*. Spikelet, X7½.

270. *Aristida intermedia* (Three Awn). *a.* Inflorescence, X½. *b.* Sheath, with ligule, X6. *c.* Spikelet, X7½.

A. *necopina* in its equal glumes and central awn of lemma longer than the lateral ones.

Aristida intermedia was collected in Illinois during September and October.

6. Aristida necopina Shinners, Rhodora 56:30. 1954. *Fig. 271.*

Strict annual; culms cespitose, to 60 cm tall; blades mostly involute, to 2 mm broad, scabrous; panicle spike-like, slender, to 25 cm long; lower glume shorter than upper, the upper 5.5–8.0 mm long, short-awned, scabrous, sharply 1-nerved; lemma 5.5–8.0 mm long, scabrous; awns divergent, essentially equal, 15–22 mm long.

COMMON NAME: Three Awn.

HABITAT: Sandy soil.

RANGE: Wisconsin, Illinois, and Indiana.

ILLINOIS DISTRIBUTION: Rare; known only from Lee County where the type was collected by V. Chase in 1935.

There may be some reason to question the specific distinctness of this species from *A. intermedia*. *Aristida necopina* has unequal glumes and equal awns of the lemma.

7. Aristida ramosissima Engelm. ex Gray, Man., ed. 2, 550. 1856.

Annual; culms branched from near the base and from most of the nodes, erect, wiry, to 50 cm tall; blades flat or involute, 1–2 mm broad; inflorescence loosely racemose, to 10 cm long; glumes 3- to 5-nerved, aristate, the first 10–20 mm long, the second 14–28 mm long; lemma 15–21 mm long; central awn stout, curved at base, 20–37 mm long; lateral awns erect or absent, up to 6 mm long.

Two forms occur in Illinois.

1. Lateral awns 0.5–6.0 mm long___7a. *A. ramosissima* f. *ramosissima*
1. Lateral awns absent, or less than 0.5 mm long_____
_____7b. *A. ramosissima* f. *uniaristata*

7a. Aristida ramosissima Engelm. f. ramosissima *Fig. 272a–c.*

Lateral awns 0.5–6.0 mm long.

271. *Aristida necopina* (Three Awn). *a*. Upper part of plants, X½. *b*. Sheath, with ligule, X6. *c*. Spikelet, X7½.

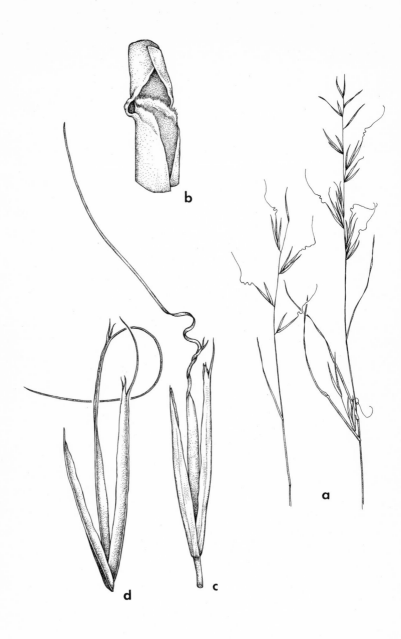

272. *Aristida ramosissima* (Slender Three Awn).—*f. ramosissima. a.* Upper part of plants, X½. *b.* Sheath, with ligule, X6. *c.* Spikelet, X3½.—*f. uniaristata. d.* Spikelet, X3½.

COMMON NAME: Slender Three Awn.
HABITAT: Dry soil of fields.
RANGE: Indiana to Iowa, south to Texas and Tennessee.
ILLINOIS DISTRIBUTION: Occasional in the southern half of the state; absent in the northern half.
The type was collected in Illinois. This species flowers during August and September. The great difference in lengths of the central and lateral awns distinguishes it.

7b. Aristida ramosissima Engelm. f. **uniaristata** (Gray) Mohlenbrock, stat. nov. *Fig. 272d.*

Aristida ramosissima var. *uniaristata* Gray, Man., ed. 5, 618. 1867.

Lateral awns absent, or less than 0.5 mm long.

HABITAT: Same as f. *ramosissima*.
RANGE: Occurring with f. *ramosissima*.
ILLINOIS DISTRIBUTION: Known from Odin, Marion County, the type locality, where it was collected by Vasey; also St. Clair County.

8. Aristida longespica Poir. in Lam. Encycl. Sup. 1:452. 1810.

Aristida gracilis Ell. Bot. S. C. & Ga. 1:142. 1816.
Annual; culms cespitose, sparsely or much branched from near the base, to 40 cm tall; blades usually involute, 1–2 mm long, confined to the lower half of the culm; inflorescence usually reduced to a very slender raceme, to 20 cm long; glumes obscurely 3-nerved, 4–9 mm long; lemma 4–6 mm long; central awn divergent, conspicuously bent near the base, 6.5–20.0 mm long; lateral awns erect to somewhat divergent, 3–15 mm long.

Two varieties occur in Illinois.

1. Lateral awns 3–4 mm long; central awn 6.5–13.0 mm long; glumes 4–6 mm long_____8a. *A. longespica* var. *longespica*
1. Lateral awns 4–15 mm long; central awn 10–20 mm long; glumes 5–9 mm long_____8b. *A. longespica* var. *geniculata*

8a. Aristida longespica Poir. var. longespica *Fig. 273.*

Lateral awns 3–4 mm long; central awn 6.5–13.0 mm long; glumes 4–6 mm long.

COMMON NAME: Three Awn.

HABITAT: Sandy soil, particularly in fields and along highways.

RANGE: Connecticut to Michigan, south to Kansas, Texas, and Florida.

ILLINOIS DISTRIBUTION: Occasional throughout the state, although rare or absent in the north-central counties. The unique feature of this species is the central awn which is conspicuously bent near the base.

8b. Aristida longespica Poir. var. geniculata (Raf.) Fern. Rhodora 35:318. 1933. *Fig. 274.*

Aristida geniculata Raf. Am. Monthly Mag. 2:119. 1817.
Lateral awns 4–15 mm long; central awn 10–20 mm long; glumes 5–9 mm long.

HABITAT: Same as var. *longespica*.

RANGE: New Hampshire to Missouri, south to Texas and Florida.

ILLINOIS DISTRIBUTION: Rare; thus far collected only in Jackson and Henry counties.

Voss (1966) believes that this variety represents the true *A. intermedia* Scribn. & Ball, a view I am rejecting at this time.

9. Aristida dichotoma Michx. Fl. Bor. Am. 1:41. 1803. *Fig. 275.*

Annual; culms cespitose, wiry, branched from near the base and at some of the nodes, erect or ascending, to 50 cm tall; blades usually involute, 1–2 mm broad; inflorescence reduced to racemes, narrow, the lowermost racemes included in the sheaths, to 8 cm long; glumes nearly equal, mucronate, 1-nerved, 5–10 mm long; lemma appressed-pubescent, 4.5–8.5 mm long; central awn divergent, coiled at the base, 3–10 mm long; lateral awns erect, straight, 0.7–3.3 mm long.

273. *Aristida longespica* (Three Awn).—var. *longespica.* *a.* Upper part of plants, X½. *b.* Sheath, with ligule, X6. *c.* Spikelet, X6.

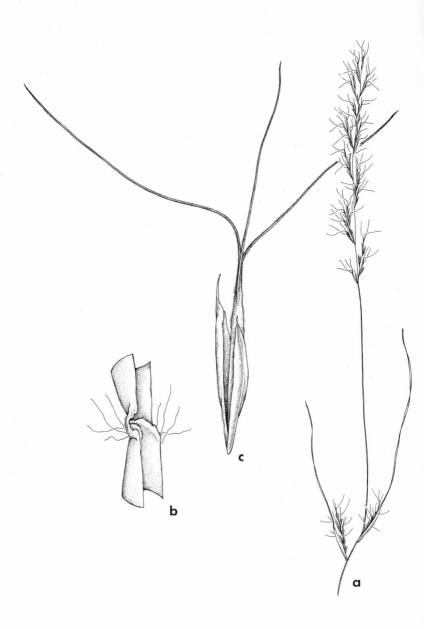

274. *Aristida longespica* (Three Awn).—var. *geniculata*, *a*. Upper part of plant, X½. *b*. Sheath, with ligule, X6. *c*. Spikelet, X6.

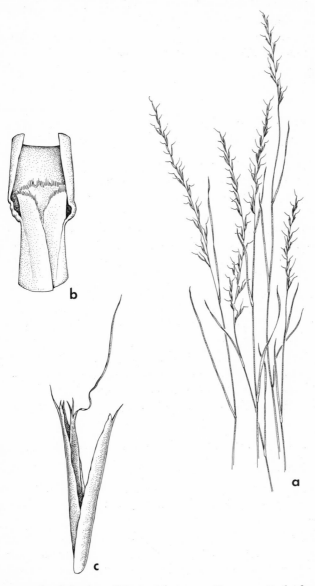

275. *Aristida dichotoma* (Three Awn).   *a.* Upper part of plants, X½. *b.* Sheath, with ligule, X6.   *c.* Spikelet, X6.

COMMON NAME: Three Awn.

HABITAT: Dry fields and along highways.

RANGE: Maine to Wisconsin, south to Kansas, Texas, and Florida.

ILLINOIS DISTRIBUTION: Occasional in the southern half of the state; less common in the northern half.

10. **Aristida basiramea** Engelm. ex Vasey, Bot. Gaz. 9:76. 1884. *Fig. 276.*

Annual; culms cespitose, branched from near the base, erect, to 60 cm tall; blades flat at base, involute at tip, about 1 mm broad; panicle slender, more or less loose, the lower included in the sheaths, to 10 cm long; glumes acuminate, 1-nerved, the first 6–12 mm long, the second 9.5–15.0 mm long; lemma 7.5–10.5 mm long; central awn divergent, coiled at base, 10–19 mm long; lateral awns erect to divergent, 5–12 mm long.

COMMON NAME: Three Awn.

HABITAT: Dry, sandy soil.

RANGE: Maine to North Dakota, south to Kansas, Tennessee, and New York.

ILLINOIS DISTRIBUTION: Not common; confined to the northern three-fourths of the state, except for single stations in Gallatin and Jackson counties.

This species flowers from August to early October.

11. **Aristida curtissii** (Gray) Nash in Britton, Man. 94. 1901. *Fig. 277.*

Aristida dichotoma var. *curtissii* Gray, Man., ed. 6, 640. 1890.
Aristida basiramea var. *curtissii* (Gray) Shinners, Am. Midl. Nat. 23:633. 1940.

Annual; culms cespitose, sparsely branched, erect or ascending, to 50 cm tall; blades usually involute, about 1 mm broad; panicles slender, sparsely branched, the lower ones included in the sheaths; first glume 5–8 mm long, the second one 7–13 mm long; lemma 7.5–9.5 mm long, scabrous on the keel; central awn divergent, coiled at base, 7.0–12.5 mm long; lateral awns erect, straight, 2–4 mm long.

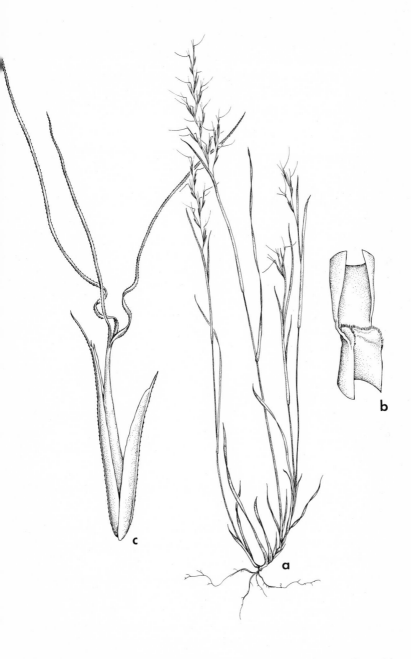

276. Aristida basiramea (Three Awn). *a.* Habit, X½. *b.* Sheath, with ligule, X6. *c.* Spikelet, X6.

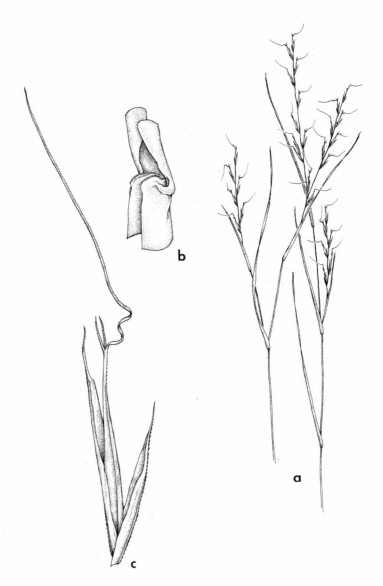

277. *Aristida curtissii* (Three Awn). *a.* Upper part of plants, X½. *b.* Sheath, with ligule, X6. *c.* Spikelet, X7½.

COMMON NAME: Three Awn.

HABITAT: Dry soil.

RANGE: New Jersey to Wyoming, south to Oklahoma and Florida.

ILLINOIS DISTRIBUTION: Rare; known only from a single collection made in Ogle County (Oregon, August 18, 1895, *W. S. Moffatt s.n.*) and one from Massac County (along CB & Q railroad at the Mermet Conservation Area, October 3, 1967, *J. Schwegman 1487*).

Although Fernald considers this taxon to be a variety of *A. dichotoma*, it appears to be related more closely to *A. basiramea*. Shinners proposed it as a variety of *A. basiramea*.

SUBFAMILY Bambusoideae

Herbaceous or woody perennials; inflorescence mostly paniculate; spikelets 1- to many-flowered; florets bisexual or unisexual.

Parodi (1961) has assigned five tribes to subfamily Bambusoideae, with only the woody Bambuseae represented in the Illinois flora.

Tribe *Bambuseae*

Usually coarse rhizomatous perennials; culms woody; leaves of two types, those on the main axis bladeless; inflorescence mostly paniculate; disarticulation above the glumes; spikelets usually large, several-flowered; glumes shorter than the lemmas; lemmas 5- to many-nerved, usually awnless; stamens usually 6; fruits various.

Only the following genus is native in Illinois.

80. *Arundinaria* MICHX. – Cane

Woody perennial from stout rhizomes; stems hollow, tall; blade of leaves petiolate, jointed to the sheath; inflorescence paniculate; spikelets many-flowered, disarticulating above the glumes; glumes very unequal, shorter than the spikelets; lemmas many-nerved; palea as long as the lemma, several-nerved, 2-keeled; lodicules 3; stamens 6; styles 2–3.

Arundinaria is one of several genera of the bamboo tribe. Species of *Phyllostachys* rarely are grown as ornamentals in southern Illinois.

Only the following species of *Arundinaria* occurs in Illinois.

278. *Arundinaria gigantea* (Giant Cane). *a.* Inflorescence, X½. *b.* Leaves, X½. *c.* Spikelet, X1. *d.* Second glume, X3. *e.* Lemma, X3.

1. **Arundinaria gigantea** (Walt.) Muhl. Cat. Pl. 14. 1813. *Fig. 278.*

Arundo gigantea Walt. Fl. Carol. 81. 1788.
Arundinaria macrosperma Michx. Fl. Bor. Am. 1:74. 1803.
Culms to 4.5 m tall, woody, hollow, erect, simple or branched, glabrous; sheaths glabrous or hirsutulous, ciliate along the margin or at the summit; ligule firm, 0.5–1.0 mm long; blades lanceolate, acute, to 35 cm long, to 3 cm broad, usually glabrous above, usually puberulent beneath, serrulate; pedicels slender, angular, 2–25 mm long; spikelets 2–5 cm long, 6- to 14-flowered, erect; glumes acuminate, pubescent, the first glume 2–6 mm long, the second glume 8–12 mm long; lemmas broadly lanceolate, keeled, 15–24 mm long, several-nerved, acuminate or with an awn to 4 mm long, ciliate, appressed-pubescent; grain ellipsoid, terete, 10–12 mm long.

COMMON NAME: Giant Cane.
HABITAT: Low woods and thickets or, in the extreme southern counties, growing near the base of rocky, wooded slopes.
RANGE: Ohio to Oklahoma, south to Texas and Florida.
ILLINOIS DISTRIBUTION: Occasional in the southern one-fourth of the state; absent elsewhere.
After the sporadically produced grains are developed, the culm bearing the spikelets dies. Spikelets are produced in April and May.

This species forms extensive "brakes" in several low areas of southern Illinois. The hollow stems are used for fishing poles and for pole bean support in rural areas. Giant Cane becomes the tallest and woodiest grass in Illinois.

Arundinaria tecta, attributed in the past to Illinois, is a more southern species with the panicles borne on shoots arising directly from the rhizomes. It apparently does not occur in Illinois.

SUBFAMILY **Oryzoideae**

Only the following tribe comprises the subfamily.

Tribe *Oryzeae*

Annuals or perennials, often with stout rhizomes; inflorescence mostly paniculate; spikelets 1-flowered, sometimes unisexual, disarticulating at base of spikelet; glumes absent; lemmas 5- to several-nerved.

Three genera of tribe Oryzeae occur in Illinois: *Leersia,*
Zizania, and *Zizaniopsis.*

81. *Leersia* sw. – Cut Grass

Perennials from scaly rhizomes; blades flat; inflorescence paniculate, composed of short racemes; spikelets 1-flowered, sometimes
sterile, disarticulating at the base; glumes none; lemma papery,
broad, keeled, 5-nerved; palea papery, nearly as long as the
lemma, keeled, usually 3-nerved.

KEY TO THE SPECIES OF Leersia IN ILLINOIS

1. Spikelets broadly rounded, 3–4 mm broad, over half as wide as
   long_____1. *L. lenticularis*
1. Spikelets oblongoid, 1–2 mm broad, less than half as wide as long.
   2. Sheaths conspicuously retrorse-scabrous; blades spinulose on
      the margins; lowest panicle branches whorled; stamens 3____
      _____2. *L. oryzoides*
   2. Sheaths glabrous or scaberulous; blades scaberulous; lowest
      panicle branches solitary; stamens 2_____3. *L. virginica*

1. Leersia lenticularis Michx. Fl. Bor. Am. 1:39. 1803. *Fig.*
279.

Asprella lenticularis (Michx.) Beauv. Ess. Agrost. 2:153. 1812.
Homalocenchrus lenticularis Kuntze, Rev. Gen. Pl. 2:777.
1891.
Perennial from scaly rhizomes; culms erect, to 1 m tall; sheaths
more or less glabrous; blades 5–20 mm broad, glabrous or softly
villous; panicle to 25 cm long, freely branched, spreading; spikelets broadly rounded, 4.0–5.5 mm long, 3–4 mm broad, over half
as wide as long, arranged in short racemes to 2 cm long; lemma
3.8–5.5 mm long, ciliate on the keel and the margins; stamens 2;
2n = 48 (Brown, 1950).

COMMON NAME: Catchfly Grass.
HABITAT: Low woodlands; swamps; marshes.
RANGE: Maryland to Minnesota, south to Texas and
Florida.
ILLINOIS DISTRIBUTION: Occasional throughout the state;
absent from the extreme northern counties.
The large, rounded spikelets readily distinguish this spe-
cies from other species of *Leersia* in Illinois. The sheaths
and blades vary from completely glabrous to pubescent.

279. *Leersia lenticularis* (Catchfly Grass). *a.* Upper part of plant, X1.
b. Sheath, with ligule, X4. *c.* Spikelet, X6.

Catchfly Grass flowers from August to October. The type was collected in Illinois by Michaux.

2. Leersia oryzoides (L.) Swartz, Prodr. Veg. Ind. Occ. 21. 1788. *Fig. 280.*

Phalaris oryzoides L. Sp. Pl. 55. 1753.
Homalocenchrus oryzoides (L.) Poll. Hist. Pl. Palat. 1:52. 1776.
Oryza clandestina f. *inclusa* Wiesb. in Baenitz. Deut. Bot. Monat. 15:19. 1897.
Leersia oryzoides f. *inclusa* (Wiesb.) Dorfl. Herb. Norm. Sched. Cent. 55–56, 164. 1915.

Perennial from slender rhizomes; culms erect or decumbent, to 1.5 m tall; sheaths conspicuously retrorse-scabrous; blades 6–12 mm broad, spinulose on the margins; panicle to 20 cm long, spreading or ascending, the lowest branches whorled; spikelets oblongoid, 3.8–6.0 mm long, 1–2 mm broad, less than half as wide as long, arranged in short racemes to 1 cm long; lemma 3.8–6.0 mm long, pilose, ciliate on the keel; stamens 3; 2n = 48 (Ramanujam, 1938).

COMMON NAME: Rice Cutgrass.
HABITAT: Low, moist soil.
RANGE: Quebec to Washington, south to California, Texas, and Florida; Europe.
ILLINOIS DISTRIBUTION: Occasional throughout the state. Anyone who has ever walked through a lowland area of Rice Cutgrass will understand the derivation of the common name. This harsh species flowers from August to mid-October. Specimens in which the panicle scarcely becomes exserted from the sheath have been called f. *inclusa.* Since the degree of exsertion seems to be mostly a matter of maturity, there is no reason to recognize this form.

3. Leersia virginica Willd. Sp. Pl. 1:325. 1797. *Fig. 281.*

Leersia ovata Poir. in Lam. Encycl. Sup. 3:329. 1813.
Homalocenchrus virginicus (Willd.) Britt. Trans. N. Y. Acad. Sci. 9:14. 1889.
Leersia virginica var. *ovata* (Poir.) Fern. Rhodora 38:386. 1936.

Perennial from short, thick rhizomes; culms erect or decumbent, rooting at the base, to 1.0 (–1.5) m tall; sheaths smooth or

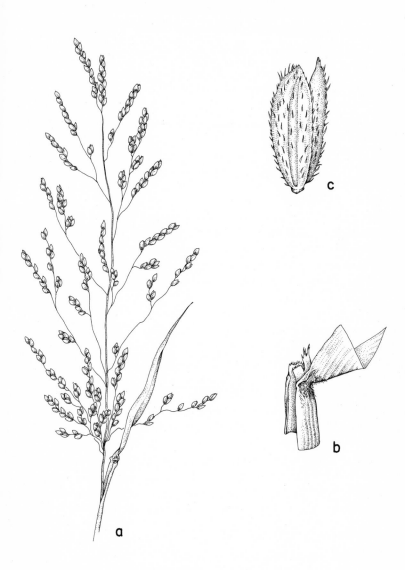

280. *Leersia oryzoides* (Rice Cutgrass). *a.* Inflorescence, X½. *b.* Sheath, with ligule, X4. *c.* Spikelet, X7½.

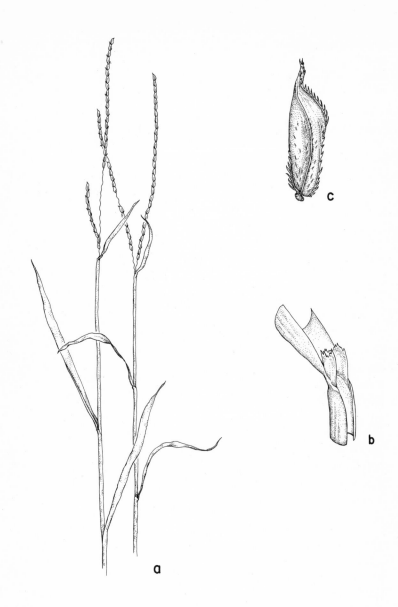

281. *Leersia virginica* (White Grass). *a.* Upper part of plants, X1.
b. Sheath, with ligule, X2½. *c.* Lemma, X7½.

scaberulous; blades 3–15 mm broad, scaberulous; panicle to 20 cm long, spreading, the lowest branches solitary; spikelets oblongoid, 2.5–4.0 mm long, 1–2 mm broad, less than half as wide as long, arranged in a short raceme to 2 cm long; lemma 3–4 mm long, glabrous to ciliate on the keel and margins; stamens 2; 2n = 24 (Brown, 1948).

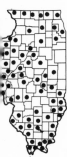

COMMON NAME: White Grass.

HABITAT: Moist woodlands.

RANGE: Quebec to South Dakota, south to Texas and Florida.

ILLINOIS DISTRIBUTION: Rather common throughout the state.

There is variation in width of the blade and in pubescence of the lemma. Some species with densely ciliate lemmas have been referred to as var. *ovata*. Since all degrees of ciliation, or lack of it, may be found in Illinois specimens and apparently correlated with no other characters, var. *ovata* is scarcely justifiable.

White Grass flowers from early July to mid-September.

82. *Zizania* L. – Wild Rice

Annuals (in Illinois); blades flat; inflorescence paniculate, large, terminal; spikelets 1-flowered, unisexual, disarticulating at the base, the pistillate appressed in concave depressions at the summit of the pedicels; glumes none; staminate lemma acuminate or short-awned, 5-nerved; pistillate lemma awned, 3-nerved.

Treatment of the North American species by Fassett (1924) is followed here. In the Hitchcock system of classification, this genus and *Zizaniopsis* were assigned to Tribe Zizanieae.

Only the following species occurs in Illinois.

1. Zizania aquatica L. Sp. Pl. 991. 1753.

Tall aquatic annual; culms slender, erect or decumbent at base, simple or branched, to nearly 3 m tall; blades 10–50 mm broad; panicle to 60 cm long, erect, the branches bearing staminate spikelets spreading to drooping, the branches bearing pistillate spikelets ascending; staminate spikelets pendulous, 6–11 mm long, glabrous or nearly so, the lemma awnless or with an awn to 3 mm long; pistillate spikelets erect, linear, or the abortive ones subulate, glabrous or scabrous, with an awn 1–6 cm long; 2n = 30 (Nandi, 1936).

Two rather difficultly distinguishable varieties are known from Illinois.

1. Pistillate lemma scabrous, slenderly nerved; aborted spikelets up to 1 mm broad, subulate, tapering into the awn_____ _____1a. Z. *aquatica* var. *aquatica*
1. Pistillate lemma glabrous, broadly nerved; aborted spikelets 1.5– 2.0 mm broad, linear, abruptly tapering into the awn_____ _____1b. Z. *aquatica* var. *interior*

1a. Zizania aquatica L. var. **aquatica** *Fig. 282a–c.*

Zizania palustris L. Mant. Pl. 295. 1771.
Blades 10–50 mm broad; pistillate lemma scabrous, slenderly nerved; aborted spikelets up to 1 mm broad, subulate, tapering into the awn.

COMMON NAME: Wild Rice.
HABITAT: Shallow water.
RANGE: Quebec to Wisconsin, south to Louisiana and Florida.
ILLINOIS DISTRIBUTION: Not common; mostly throughout the state, except the southern counties.
This is one of the tallest grasses in Illinois. It can be confused with no other species in Illinois because of its large panicles with unisexual spikelets borne on the same branchlets.
It flowers from June to September.

1b. Zizania aquatica L. var. **interior** Fassett, Rhodora 26:158. 1924. *Fig. 282d.*

Zizania interior (Fassett) Rydb. Brittonia 1:82. 1931.
Blades 10–30 mm broad; pistillate lemma glabrous, broadly nerved; aborted spikelets 1.5–2.0 mm broad, linear, abruptly tapering into the awn.

COMMON NAME: Wild Rice.
HABITAT: Shallow water.
RANGE: Indiana to North Dakota, south to Texas and Missouri.
ILLINOIS DISTRIBUTION: Known from Cook, Lake, and Union counties. The completely smooth pistillate lemma is about the most reliable character in separating this variety from var. *aquatica*.

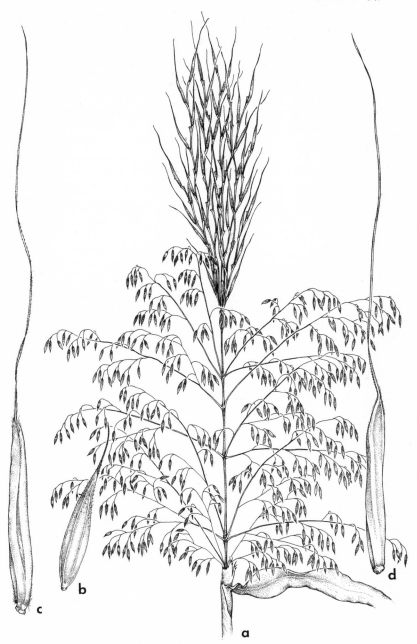

282. *Zizania aquatica* (Wild Rice).—var. *aquatica*. *a.* Inflorescence, X½.
b. Staminate lemma, X7½. *c.* Pistillate lemma, X7½.—var. *interior*. *d.* Pistillate lemma, X7½.

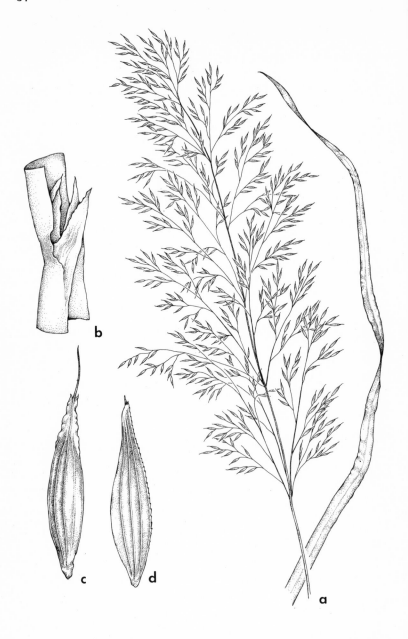

283. *Zizaniopsis miliacea* (Southern Wild Rice). *a.* Inflorescence, X½.
b. Sheath, with ligule, X5. *c.* Pistillate lemma, X3. *d.* Staminate lemma,
X3.

83. *Zizaniopsis* DOELL & ASCHERS

Large, succulent perennials from stout rhizomes; inflorescence paniculate; spikelets 1-flowered, unisexual, with staminate flowers below and pistillate flowers above on same branches; disarticulation below the spikelet; glumes absent; lemma 5-nerved and awnless in the staminate spikelet, 7-nerved and short-awned in the pistillate spikelet; stamens 6.

Only the following species occurs in Illinois.

1. Zizaniopsis miliacea (Michx.) Doell & Aschers in Doell in
Mart. Fl. Bras. 2(2):13. 1871. *Fig. 283.*

Zizania miliacea Michx. Fl. Bor. Am. 1:74. 1803.
Robust perennial; culms to 3 m tall, glabrous; leaves to 2 cm broad, with very rough margins; panicle nodding, to 60 cm long, with numerous, whorled branches; lemmas 6–9 mm long, with scabrous nerves, the pistillate ones with an awn to 5 mm long.

COMMON NAME: Southern Wild Rice.
HABITAT: Edge of lake (in Illinois).
RANGE: Maryland to Kentucky, southeast Missouri, and eastern Oklahoma, south to Texas and Florida.
ILLINOIS DISTRIBUTION: Montgomery County (shallow water at edge of Lake Hillsboro, October 4, 1961, *G. S. Winterringer 18526*). This robust southern species occurs in several clumps in shallow water along the shore of Lake Hillsboro. It is not known whether the species is native at this location, but a local resident has indicated its presence in the lake for nearly thirty years.

SUBFAMILY **Arundinoideae**

Annuals or perennials; inflorescence paniculate or racemose; spikelets 1- to several-flowered; disarticulation above the glumes.

The three tribes included in this subfamily by Gould (1968) are composed of grasses quite unlike in appearance. It is unfortunate that the name of the subfamily is so similar to *Arundinaria* since this genus is not assigned to this subfamily. The derivation for the name of the subfamily is from the ornamental genus *Arundo*.

Tribe *Arundineae*

Coarse perennials; inflorescence paniculate; spikelet more than

1-flowered; disarticulation above the glumes; glumes thin; lemmas usually 3-nerved.

Only the following genus occurs naturally in Illinois.

84. Phragmites TRIN. – Reed

Very tall perennials from stout, creeping rhizomes; blades broad, flat; inflorescence paniculate, large, dense, much-branched; spikelets several-flowered, disarticulating above the glumes, the rachilla silky-villous; glumes 2, unequal; lemmas glabrous, 3-nerved, the lowest sterile or bearing a staminate flower, the others perfect.

A closely related tall grass, the Giant Reed (*Arundo donax* L.), is cultivated occasionally as an ornamental in Illinois. It differs from *Phragmites* by having pubescent lemmas and a glabrous rachilla.

Only the following species of *Phragmites* occurs in Illinois.

1. Phragmites australis (Cav.) Trin. ex Steud. Nom. Bot. ed.

2, 2:324. 1841. *Fig. 284.*

Arundo phragmites L. Sp. Pl. 81. 1753.
Arundo australis Cav. Ann. Hist. Nat. 1:100. 1799.
Phragmites communis Trin. Fund. Agrost. 134. 1820.
Phragmites berlandieri Fourn. Bull. Bot. Soc. France 24:178. 1877.
Phragmites communis var. *berlandieri* (Fourn.) Fern. Rhodora 34:211. 1932.

Tall perennial with culms to 4 m high, usually forming extensive colonies; blades flat, glabrous, 10–50 mm broad; inflorescence large, 15–40 cm long, mostly ascending, yellowish to purplish; spikelets 3- to 7-flowered, 10–17 mm long, surpassed by the silky hairs of the rachilla; first glume narrowly elliptic, obtuse to subacute, 4–6 mm long; second glume linear, acute, 6.0–8.5 mm long; lemmas linear-lanceolate, glabrous, 3-nerved, the lowest 8–12 mm long, the upper progressively smaller, with all lemmas attaining the same height; 2n = 48 or 96 (Avdulov, 1931).

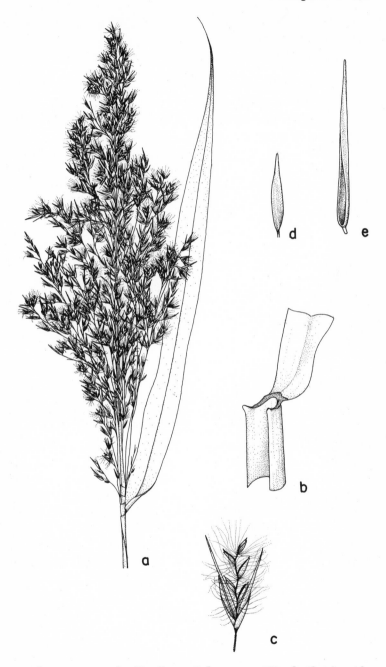

284. *Phragmites australis* (Reed). *a.* Inflorescence, X½. *b.* Sheath, with ligule, X1. *c.* Spikelet, X2. *d.* First glume, X4. *e.* Second lemma, X4.

COMMON NAME: Reed.

HABITAT: Moist soil.

RANGE: Throughout North America; South America; Europe; Asia; Africa; Australia.

ILLINOIS DISTRIBUTION: Occasional in the northern half of the state, but uncommon in the southern half. This is one of the tallest native grasses in Illinois. The long, silky hairs of the rachilla make this a most handsome species. The flowers are produced from late July to late September.

Clayton (1968) has given sound evidence why this species should be called *P. australis* rather than *P. communis.*

Tribe *Centotheceae*

Perennials; inflorescence paniculate; spikelets several-flowered; disarticulation above the glumes; glumes shorter than the lemmas; lemmas several-nerved.

Only the following genus occurs in Illinois.

85. Chasmanthium LINK – Sea Oats

Tall rhizomatous perennials; leaves flat (in Illinois specimens) or involute; inflorescence paniculate; spikelets 3- to many-flowered, compressed, disarticulating above the glumes; glumes 2, keeled, unequal, several-nerved; lemmas chartaceous, many-nerved, the lowest 1 (in Illinois) –5 lemmas sterile; stamen 1.

This genus is unique among most grasses in having the lower 1–5 lemmas sterile. The large, compressed spikelets make this a handsome grass for cultivation.

The species placed in this genus is more familiarly known as *Uniola latifolia* Michx. Yates (1966a, 1966b, 1966c) has presented evidence that this species, along with four others previously assigned to *Uniola,* are different in several respects from other species placed in *Uniola.* In fact, according to Yates, *Uniola* and the segregated *Chasmanthium* are now referred not only to different tribes but also to different subfamilies!

Only the following species occurs in Illinois.

1. Chasmanthium latifolium (Michx.) Yates, Southw. Nat. 11:416. 1966. *Fig. 285.*

Uniola latifolia Michx. Fl. Bor. Am. 1:70. 1803.

Perennial from short, thick rhizomes; culms to 1.5 m tall; blades

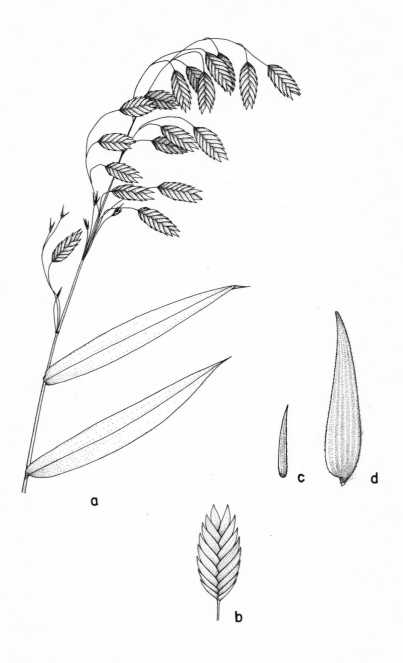

285. Chasmanthium latifolium (Sea Oats).   *a.* Upper part of plant, X½.
b. Spikelet, X1.   *c.* Glume, X4.   *d.* Second fertile lemma, X4.

flat, 8–25 mm broad, glabrous; inflorescence 10–20 cm long, nodding; spikelets compressed, 15–40 mm long, up to 20 mm broad, 6- to 18-flowered, green at first, becoming brownish; glumes keeled, linear-lanceolate, acute, 3–8 mm long; lowest lemma sterile, keeled, linear-lanceolate, acute, 3–9 mm long; fertile lemmas keeled, lanceolate, acute, 4–10 mm long, hispidulous along the keel; 2n = 24 (Avdulov, 1931).

COMMON NAME: Sea Oats; Wild Oats.

HABITAT: Moist soil.

RANGE: Pennsylvania to Kansas, south to Arizona and Florida.

ILLINOIS DISTRIBUTION: Common in the southern half of the state, rare in the northern half.

The long-persistent spikelets make this an appealing plant for winter bouquets. The flowers bloom from July to early October.

The Michaux type is reported by Hitchcock (1935) to be from Illinois.

Tribe *Danthonieae*

Annuals or perennials; inflorescence mostly paniculate or racemose; spikelets several-flowered; disarticulation above the glumes; glumes equaling or longer than the lemmas; lemmas several-nerved, notched at the apex.

The segregation of *Danthonia* and a few other primarily Southern Hemisphere genera into a separate tribe is explained by Hubbard (1948).

Only the following genus occurs in Illinois.

86. *Danthonia* LAM. & DC. – Wild Oat Grass

Perennials; blades mostly involute; inflorescence paniculate (appearing racemose in the Illinois species); spikelets 3- to 6-flowered, disarticulating above the glumes; glumes subequal, longer than the lemmas; lemmas rounded on the back, pubescent, with two apical teeth with an awn arising between the teeth.

Only the following species occurs in Illinois.

1. **Danthonia spicata** (L.) Beauv. ex Roem. & Schult. Syst. Veg. 2:690. 1817. *Fig. 286.*

Avena spicata L. Sp. Pl. 80. 1753.

Densely cespitose perennial with glabrous or pilose culms to 50

286. *Danthonia spicata* (Poverty Oat Grass). *a*. Upper part of plants, X½.
b. Sheath, with ligule, X2½. *c*. Spikelet, X5. *d*. Glumes, X6. *e*. Lemma,
X6.

cm tall; sheaths glabrous or pilose; blades mostly involute, 1–2 mm broad, glabrous or pilose, becoming curled when old; panicle contracted, 2–5 cm long, appearing racemose; spikelets 8–15 mm long; glumes lanceolate, 3-nerved, acuminate, thin, 8–15 mm long, pubescent; lemmas ovate, acute, pilose, 3.5–5.0 mm long; awn 4.5–7.0 mm long, seldom straight.

COMMON NAME: Poverty Oat Grass; Curly Grass.
HABITAT: Dry woods and bluffs.
RANGE: Newfoundland to British Columbia, south to New Mexico and Florida.
ILLINOIS DISTRIBUTION: Occasional to common throughout the state.
Variety *longipila* Scribn. & Merr., whose range includes Illinois, has not been recorded as yet from this state. It is more slender than var. *spicata* and has the florets of each spikelet much shorter than the glumes.
Poverty Oat Grass flowers from late May through August.

Species Excluded

Aristida pallens Nutt. Mead's report (1846) of this species from Illinois is an error for *A. oligantha.*

Arundinaria tecta (Walt.) Muhl. The several reports of this species from Illinois are misidentifications for *A. gigantea.*

Cenchrus carolinianus Walt. Mosher (1918) and others used this binomial for *Cenchrus longispinus,* but these two are completely different species.

Cenchrus pauciflorus Benth. This binomial has been consistently applied to the Sand Bur in Illinois by workers since 1950. It is, however, a name for a species unknown from Illinois.

Cenchrus tribuloides L. From 1846 to the end of the nineteenth century, the Illinois Sand Bur was called *C. tribuloides.* This latter binomial does not apply to any Illinois species.

Erianthus contortus Baldw. ex Ell. This is a Coastal Plain species which Lapham (1857) erroneously attributed to Illinois.

Erianthus saccharoides Michx. Brendel (1887) mistakenly identified *E. alopecuroides* from Illinois as *E. saccharoides.*

Muhlenbergia richardsonis (Trin.) Rydb. Fassett (1927) reported this western species from Illinois on the basis of a collection by J. Wolf.

Muhlenbergia foliosa (R. & S.) Trin. During the period when *M. frondosa* was erroneously called *M. mexicana,* the true *M. mexicana* was referred to as *M. foliosa.* This latter binomial, however, applies to a different species.

Panicum ovale Ell. Jones et al. (1955) attribute a V. H. Chase collection from Mason County to this species. In their addenda, however, they exclude this species on the advice of Mrs. Agnes Chase.

Panicum proliferum Lam. This binomial was used for *P. dichotomiflorum* Michx. by Mead (1846), Lapham (1857), Patterson (1874, 1876), and others. *Panicum proliferum* is a different species, however.

Panicum pubescens Lam. Engelmann (1843) and later Huett (1898) used this binomial for *P. lanuginosum* var. *fasciculatum.* A specimen in the Mead herbarium labelled *P. pubescens* Lam. is actually *P. leibergii. Panicum pubescens* is a different species.

Panicum xanthophysum Gray. Lapham (1857) erroneously

attributed this binomial to Illinois on plants which actually are
P. leibergii.

Poa hirsuta (Michx.) Nees. This is the binomial which Engelmann (1843) and Mead (1846) erroneously called specimens of
Eragrostis spectabilis.

Spartina cynosuroides (L.) Roth. Until the first decade of the
twentieth century, many Illinois botanists used this binomial for
S. pectinata. Spartina cynosuroides is an entirely different species, however.

Spartina polystachya (Michx.) Beauv. Short's (1845) mis-
identification of *S. pectinata* was called *S. polystachya.*

Sporobolus junceus (Michx.) Kunth. Although this species
does not occur in Illinois, it was attributed to this state by
Higley and Raddin (1891) and Pepoon (1927) based on mis-
identified specimens of *S. heterolepis.*

Sporobolus virginicus L. Pepoon's report (1927) of this species from Lake County is an error for *Agrostis alba* var. *palustris.*

Vilfa virginica (L.) Beauv. Early workers confused this species, which does not occur in Illinois, with *Sporobolus neglectus.*

Summary of the Taxa of Grasses in Illinois

	Genera	Species	Lesser Taxa
Festucoideae	(40)	(120)	(18)
Festuceae	8	49	2
Aveneae	18	32	4
Triticeae	7	22	11
Meliceae	3	9	1
Stipeae	2	6	0
Brachyelytreae	1	1	0
Diarrheneae	1	1	0
Panicoideae	(19)	(89)	(23)
Paniceae	9	71	20
Andropogoneae	10	18	3
Eragrostoideae	(20)	(68)	(7)
Eragrosteae	8	39	5
Chlorideae	10	17	0
Aeluropodeae	1	1	0
Aristideae	1	11	2
Bambusoideae	(1)	(1)	(0)
Bambuseae	1	1	0
Oryzoideae	(3)	(5)	(1)
Oryzeae	3	5	1
Arundinoideae	(3)	(3)	(0)
Arundineae	1	1	0
Danthonieae	1	1	0
Centotheceae	1	1	0
Totals	86	286	49

GLOSSARY
LITERATURE CITED
INDEX OF PLANT NAMES

Acuminate. Gradually tapering to an elongated point.

Acute. Sharp, ending in a point.

Annual. Living for a single year.

Anther. The terminal part of a stamen which bears the pollen.

Antrorse. Pointing upward.

Apiculate. Ending abruptly in a small, sharp tip.

Appressed. Lying flat against the surface.

Aristate. Bearing an awn.

Attenuate. Long-tapering.

Auriculate. Bearing an ear-like process.

Awn. A bristle usually terminating a structure.

Axis. The central support to which lateral parts are attached.

Bidentate. Having two teeth.

Bifid. Two-cleft.

Callus. A hard swollen area at the outside base of a lemma or palea.

Canescent. Grayish-hairy.

Capillary. Threadlike.

Carinate. Bearing a keel.

Cartilaginous. Firm but flexible.

Caryopsis. A type of one-seeded, dry, indehiscent fruit with seed coat attached to the mature ovary wall.

Caudex. (pl., **caudices**). The woody base of a perennial plant.

Cauline. Belonging to a stem.

Cespitose. Growing in a tuft.

Chartaceous. Papery.

Ciliate. Bearing marginal hairs.

Compressed. Flattened.

Conduplicate. Folded together lengthwise.

Connate. United, when referring to like parts.

Connivent. Coming in contact; converging.

Convex. Rounded on the outer surface; opposite of concave.

Coriaceous. Leathery.

Culm. The stem which terminates in an inflorescence.

Cuspidate. Terminating in a very short point.

Decumbent. Lying flat, but with the tip ascending.

Diffuse. Loosely spreading.

Digitate. Radiating from a common point, like the finger from a hand.

Dioecious. With staminate flowers on one plant, pistillate flowers on another.

Disarticulate. To come apart; to become disjointed.

Divergent. Spreading apart.

Ellipsoid. Referring to a solid object which, in side view, is broadest at the middle, gradually tapering equally to both ends.

Elliptic. Broadest at the middle, gradually tapering equally to both ends.

Emarginate. Deeply notched at the tip.

Erose. With an irregularly notched margin.

Fascicle. A cluster; a bundle.

Fibrous. Bearing fibers; i.e., slender projections of equal diameters.

Filiform. Threadlike.

Flexuous. Zigzag.

Floret. A small flower.

Geniculate. Bent.

Glabrate. Becoming smooth.

Glabrous. Smooth; without hairs, scales, or glands.

Glaucous. With a whitish covering which can be rubbed off.

Glume. A sterile scale subtending a spikelet.

Grain. The fruit of most grasses.

Hispid. With rigid hairs.

Hispidulous. With minute rigid hairs.

Hirsute. With stiff hairs.

Hirtellous. With minute stiff hairs.

Hyaline. Transparent.

Indurate. Hardened.

Inflorescence. A cluster of flowers.

Internode. The area between two consecutive nodes.

Involute. Rolled inward.

Keel. A central ridge.

Lanceolate. Lance-shaped; broadest near base, gradually tapering to the narrow apex.

Lanceoloid. Referring to a solid object which is broadest near base, gradually tapering to the narrow apex.

Lemma. A scale subtending the floret.

Ligule. The structure on the inner surface of the leaf at the junction of the blade and the sheath.

Linear. Elongated and uniform in width throughout.

Lodicule. A small rudimentary structure at the base of a grass flower.

Monoecious. With stamens and pistils in separate flowers on the same plant.

Mucronate. Bearing a short, terminal point.

Nerve. Vein.

Node. That place on the stem from which leaves and branchlets arise.

Oblong. With nearly uniform width throughout, but broader than linear.

Oblongoid. Referring to a solid object which, in side view, is nearly uniform in width throughout.

Obovate. Broadly rounded at apex, becoming narrowed below; broader than oblanceolate.

Obsolete. Not apparent.

Obtuse. Rounded; blunt.

Orbicular. Round.

Ovary. The lower swollen part of the pistil which produces the ovules.

Ovoid. Referring to a solid object which, in side view, is broadly rounded at base, becoming narrowed above.

Ovule. The egg-producing structure found within the ovary; an immature seed.

Palea. The scale opposite the lemma which encloses the flower.

Panicle. A type of inflorescence composed of several racemes.

Papillose. Bearing pimple-like processes.

Pedicel. The individual stalk of a spikelet.

Pedicellate. Bearing a pedicel.

Peduncle. The stalk of an inflorescence.

Perennial. Living more than one year.

Perfect. Bearing both stamens and pistils.

Perianth. That part of the flower composed of the calyx or corolla or both.

Pericarp. The ripened ovary wall.

Pilose. Bearing soft long hairs.

Pistil. Female reproductive organ.

Plicate. Folded.

Prostrate. Lying flat.

Puberulent. Minutely pubescent.

Raceme. A type of inflorescence where pedicellate flowers are arranged along an elongated axis.

Racemose. Bearing racemes.

Rachilla. The axis bearing the flowers.

Rank. Referring to the number of planes in which structures are borne.

Reflexed. Turned downward.

Retrorse. Pointing downward.

Retuse. Shallowly notched at a rounded apex.

Rhizomatous. Bearing rhizomes.

Rugose. Wrinkled.

Rugulose. With small wrinkles.

Scaberulous. Slightly rough to the touch.

Scabrous. Rough to the touch.

Scarious. Thin and membranous.

Sericeous. Silky; bearing soft, appressed hairs.

Serrate. With teeth which project forward.

Serrulate. With very small teeth which project forward.

Sessile. Without a stalk.

Seta. Bristle.

Setose. Bearing setae.

Setulose. Bearing small setae.

Sheath. A protective covering; the basal part of a grass leaf that encircles the stem.

Spicate. Bearing a spike.

Spike. A type of inflorescence where sessile flowers are arranged along an elongated axis.

Spikelet. The basic unit in a grass inflorescence.

Spinulose. With small spines.

Stamen. The male reproductive organ.

Staminate. Bearing stamens.

Stigma. The apex of the pistil which receives the pollen.

Stipitate. Bearing a stipe or stalk.

Stolon. A slender, horizontal stem on the surface of the ground.

Stoloniferous. Bearing stolons.

Strigose. With appressed, straight hairs.

Style. That elongated part of the pistil between the ovary and the stigma.

Subulate. With a very short, narrow point.

Terete. Round in cross section.

Translucent. Partly transparent.

Truncate. Abruptly cut across.

Umbonate. With a stout projection at the center.

Villous. With long, soft, slender, unmatted hairs.

Viscid. Sticky.

Whorled. An arrangement of three or more structures at a point on the stem.

LITERATURE CITED

Ali, M. A. 1967. The *Echinochloa crusgalli* complex in the United States. Unpublished Ph.D. dissertation, Texas A. & M. University.

Avdulov, N. P. 1928. Karyo-systematische Untersuchungen der Familie Gramineen. All Union Cong. Bot. Moscow Jour. 65–67.

———. 1931. Karyo-systematische Untersuchungen der Familie Gramineen. Bulletin of Applied Botany, Genetics, and Plant Breeding, Suppl. 44. 428 pp.

Beetle, A. A. 1943. The North American variations of *Distichlis spicata*. Bulletin of the Torrey Botanical Club 70:638–50.

Brendel, F. 1887. Flora Peoriana. Peoria, Illinois.

Brown, W. V. 1948. A cytological study in the Gramineae. American Journal of Botany 35:382–95.

———. 1950. A cytological study of some Texas Gramineae. Bulletin of the Torrey Botanical Club 77:63–76.

———. 1958. Leaf anatomy in grass systematics. Botanical Gazette 119:170–78.

Chase, A. 1951. Hitchcock's Manual of the Grasses of the United States. 2nd ed. rev. United States Department of Agriculture Miscellaneous Publication 200. 1051 pp.

Church, G. 1929. Meiotic phenomena in certain Gramineae. I. Festuceae, Aveneae, Agrostideae, Chlorideae, and Phalarideae. Botanical Gazette 87:608–29.

———. 1929a. Meiotic phenomena in certain Gramineae. II. Paniceae and Andropogoneae. Botanical Gazette 88:63–84.

———. 1936. Cytological studies in the Gramineae. American Journal of Botany 23:12–15.

Clayton, W. D. 1968. The correct name of the common reed. Taxon 17:168–69.

Darlington, C. D. and M. B. Upcott. 1941. The activity of inert chromosomes in *Zea mays*. Journal of Genetics 41:275–95.

DeLisle, D. G. 1963. Taxonomy and distribution of the genus *Cenchrus*. Iowa State College Journal of Science 37:259–351.

Engelmann, G. 1843. Catalogue of collection of plants made in Illinois and Missouri, by Charles A. Geyer. American Journal of Science 46:94–104.

Evers, R. A. 1949. *Setaria faberii* in Illinois. Rhodora 51:391–92.

Fassett, N. C. 1924. A study of the genus *Zizania*. Rhodora 26:153–60.

———. 1927. Notes from the herbarium of the University of Wisconsin. I. Rhodora 29:227–34.

Fassett, N. C. 1949. Some notes on *Echinochloa*. Rhodora 51:1–3.

Fernald, M. L. 1943. Five common rhizomatous species of *Muhlenbergia*. Rhodora 45:221–39.

———. 1945. An incomplete flora of Illinois. Rhodora 47:204–19.

———. 1950. Gray's Manual of Botany. 8th ed. New York: The American Book Co. 1632 pp.

Fults, J. L. 1942. Somatic chromosome complements in *Bouteloua*. American Journal of Botany 29:45–55.

Glassman, S. F. 1964. Grass flora of the Chicago region. The American Midland Naturalist 72:1–49.

Gleason, H. A. 1910. The vegetation of the inland sand deposits of Illinois. Bulletin of the Illinois State Laboratory of Natural History 9:23–174.

———. 1952. The New Britton and Brown Illustrated Flora of the Northeastern United States and Adjacent Canada. I. New York: The New York Botanical Garden.

Gould, F. 1967. The grass genus *Andropogon* in the United States. Brittonia 19:68–73.

———. 1968. Grass Systematics. New York: McGraw-Hill Book Company. 382 pp.

Harvey, L. H. 1948. *Eragrostis* in North and Middle America. Unpublished Ph.D. dissertation, University of Michigan.

Henrard, J. T. 1927. A critical revision of the genus *Aristida*. Meded. Van Rijks Herb. 54:1–220.

Higley, W. K. and C. S. Raddin. 1891. Flora of Cook County, Illinois and a part of Lake County, Indiana. Bulletin of the Chicago Academy of Sciences 2:1–168.

Hitchcock, A. S. 1920. The Genera of Grasses of the United States, with special reference to the economic species. United States Department of Agriculture Bulletin 772.

———. 1935. Manual of the Grasses of the United States. United States Department of Agriculture Miscellaneous Publication Number 200.

——— and A. Chase. 1910. The North American species of *Panicum*. Contributions from the U.S. National Herbarium 15:1–396.

Hubbard, C. E. 1948. The genera of British grasses. In J. Hutchinson, British Flowering Plants, pp. 284–348.

Huett, J. W. 1898. Essay toward a natural history of LaSalle County, Illinois. Flora LaSallensis, part 2.

Hunter, A. W. 1934. A karyosystematic investigation in the Gramineae. Canadian Journal of Research, C, 11:213–41.

Huskins, C. L. and S. G. Smith. 1934. A cytological study of the genus *Sorghum* Pers. 2. The meiotic chromosomes. Journal of Genetics 28:387–95.

Hutchinson, J. 1959. The Families of Flowering Plants. Vol. 2. Monocotyledons. Oxford: Clarendon Press.

Jones, G. N. 1945. Flora of Illinois. 1 ed. South Bend: University of Notre Dame Press. 317 pp.
————. 1950. Flora of Illinois. 2 ed. South Bend: University of Notre Dame Press. 368 pp.
———— et al. 1955. Vascular Plants of Illinois. Urbana: University of Illinois Press, and the Illinois State Museum, Springfield. 593 pp.

Kibbe, A. 1952. A botanical study and survey of a typical mid-western county (Hancock County, Illinois). Carthage, Illinois. 425 pp.

Kishimoto, E. 1938. Chromosomenzahlen in den Gattungen Panicum and Setaria. I. Chromosomenzahlen einiger Setaria-Arten. Cytologia 9:23–27.

Lapham, I. A. 1857. The native, naturalized, and cultivated grasses of the state of Illinois. Transactions of the Illinois State Agricultural Society 2:551–613.

Lorch, J. 1962. A revision of Crypsis Ait. (Gramineae). Bulletin of the Research Council of Israel 11D:91–102.

Mangelsdorf, P. C. and R. G. Reeves. 1939. The origin of Indian corn and its relatives. Texas Agriculture Experimental Station Bulletin no. 574.

Mead, S. B. 1846. Catalogue of plants growing spontaneously in the state of Illinois, the principal part near Augusta, Hancock County. Prairie Farmer 6:35–36; 60; 93; 119–22.

Mobberley, D. G. 1956. Taxonomy and distribution of the genus Spartina. Iowa State College Journal of Science 30:471–574.

Moffett, A. A. and R. Hurcombe. 1949. Chromosome numbers of South African grasses. Heredity 3:369–73.

Mosher, E. 1918. The grasses of Illinois. Bulletin of the Agricultural Experiment Station, University of Illinois 205:257–425.

Mukherjee, S. K. 1958. Revision of the genus Erianthus Michx. (Gramineae). Lloydia 21:157–88.

Nandi, H. K. 1936. The chromosome homology, secondary association and origin of cultivated rice. Journal of Genetics 33:315–36.

Nicora, E. G. 1962. Revalidacion del género de Gramineas Neeragrostis de la flora Norte americana. Rev. Argent. Agron. 29:1–11.

Nielsen, E. L. 1939. Grass Studies. III. Additional somatic chromosome complements. American Journal of Botany 26:366–72.

Parodi, L. R. 1943. Una neuva especie de "sorghum," cultivada en la Argentina. Revista Argentina de Agronomía 10:361–72.

————. 1961. La taxonomia de las Gramineae Argentinas a la luz de las investigaciones mas recentes. Recent Advances in Botany 1:125–29.

Patterson, H. N. 1874. A list of plants collected in the vicinity of Oquawka, Henderson County. Oquawka, Illinois. 18 pp.

————. 1876. Catalogue of the phaenogamous and vascular cryptogamous plants of Illinois. Oquawka, Illinois. 54 pp.

Pepoon, H. S. 1927. An annotated flora of the Chicago area. Bulletin of the Chicago Academy of Sciences 8:1–554.

Pohl, R. W. 1969. *Muhlenbergia*, Subgenus Muhlenbergia (Gramineae) in North America. American Midland Naturalist 82:512–42.

Ramanujam, S. 1938. Cytogenetical studies in the Oryzeae. I. Chromosome studies in the Oryzeae. Annals of Botany, London, N.S. 2:107–25.

Randolph, L. F. 1932. Some effects of high temperature on polyploidy and other variations in maize. Proceedings of the National Academy of Science 18:222–29.

Reeder, J. R. 1951. *Setaria lutescens*, an untenable name. Rhodora 53:27–30.

———. 1957. The embryo in grass systematics. American Journal of Botany 44:756–68.

——— and M. A. Ellington. 1960. *Calamovilfa:* A misplaced genus of Gramineae. Brittonia 12:71–77.

Rominger, J. M. 1962. Taxonomy of *Setaria* (Gramineae) in North America. Illinois Biological Monographs no. 29. Urbana: University of Illinois Press.

Saura, F. 1943. Cariología de Gramíneas: géneros *Paspalum, Stipa, Poa, Andropogon y Phalaris.* Rev. Fac. Agron. Buenos Aires 10:344–53.

Short, C. W. 1845. Observations on the Botany of Illinois. Western Journal of Medicine and Surgery 3:185–98.

Snyder, L. A. and J. R. Harlan. 1953. A cytological study of *Bouteloua gracilis* from western Texas and eastern New Mexico. American Journal of Botany 40:702–7.

Stebbins, G. L. and B. Crampton. 1961. A suggested revision of the grass genera of temperate North America. Recent Advances in Botany 1:133–45.

——— and R. M. Love. 1941. A cytological study of California forage grasses. American Journal of Botany 28:371–82.

Thieret, J. W. 1960. *Calamovilfa longifolia* and its variety *magna.* American Midland Naturalist 63(1):169–76.

Thorne, R. F. 1968. Synopsis of a putatively phylogenetic classification of the flowering plants. Aliso 6:57–66.

Voigt, J. W. 1951. Additional collections of *Andropogon elliottii* Chapm. in southern Illinois. Rhodora 53:128–30.

———. 1953. Yield and consumption in a southern Illinois bluegrass-broom sedge pasture. Journal of Range Management 6:260–66.

Voss, E. G. 1966. Nomenclatural notes on monocots. Rhodora 68:435–63.

Wiegand, K. 1921. The genus *Echinochloa* in North America. Rhodora 23:49–65.

Winterringer, G. S. and R. A. Evers. 1960. New Records for Illinois Vascular Plants. Scientific Papers Series. 11. The Illinois State Museum, Springfield.

Yates, H. O. 1966a. Morphology and cytology of *Uniola* (Gramineae). Southwestern Naturalist 11:145–89.

————. 1966b. Revision of grasses traditionally referred to *Uniola*. I. *Uniola* and *Leptochloöpsis*. Southwestern Naturalist 11:372–94.

————. 1966c. Revision of grasses traditionally referred to *Uniola*. II. *Chasmanthium*. Southwestern Naturalist 11:415–55.

INDEX OF PLANT NAMES